Mentals 3

Alan McSeveny Rachel McSeveny Diane McSeveny-Foster

Pearson Australia
(a division of Pearson Australia Group Pty Ltd)
459-471 Church St, Level 1, Building B, Richmond, Victoria, 3121
PO Box 23360, Melbourne, Victoria 8012
www.pearson.com.au

First published 2024 by Pearson Australia
2028 2027 2026 2025
10 9 8 7 6 5 4 3 2 1

Publishers: Sophie Matta and Kerry Nagle
Project Manager: Michelle Thomas
Production Editor: Laura Rentsch
Editor: Rachel Elliott
Designer: Anne Donald
Proofreader: Ann M. Philpott
Rights & Permissions Editor: Alice McBroom
Cover art: Michael Barter
Illustrator: Michael Barter
Publishing Services: Jit-Pin Chong
Printed in Malaysia by Vivar

ISBN 978 0 655 708 834
Pearson Australia Group Pty Ltd ABN 40 004 245 943

Attributions
We would like to thank the following for permission to reproduce copyright material.

Shutterstock: NYgraphic (rose), p. 67.

Acknowledgement of Country
Pearson respects and honours Aboriginal and Torres Strait Islander Elders past, present and future. We acknowledge the stories, traditions and living cultures of the Traditional Custodians of the lands on which our company is located and where we conduct our business. Pearson is committed to honouring Australian Aboriginal and Torres Strait Islander peoples' unique cultural and spiritual relationships to the land, waters and seas and their rich contribution to society.

Aboriginal and Torres Strait Islander peoples are advised that this text may contain images, voices and names of deceased persons.

Introduction

Using the Mentals Books

This book is used most effectively when it aligns with the suggested program in the Student Book contents.

Each unit of the Mentals Book is programmed to review Student Book content for the previous two weeks. For example, Unit 15 of the Mentals Book can be set as homework to review weeks 13 and 14 of the Student Book while week 15 of the Student Book is being taught.

Units 1 and 2 of this book revise work from the previous year and could be completed in weeks 1 and 2 of the school year.

Mixed-topic questions

The units present questions in a mixed-topic format to encourage thorough understanding and continuous review.

Presentation

- Number facts are reinforced to encourage instant recall.
- Essential skills are explained.
- The Arithmetic Card (page 5) is a useful teaching tool for practising basic number skills.
- ID cards (pages 6 and 7) review the mathematical terms students need to learn.
- Measurement standards and examples (pages 8 and 9) are provided so that students can learn important facts and estimate measurements effectively.

Graded questions

- Column 1: easier
- Column 2 and 3: harder
- Column 4: Extension and Challenge

Motivation

- There are two lizards hidden on each page for students to find.
- The header allows students to record their score.

Extra activities

- Problem-solving **strategies** are introduced in a carefully planned sequence throughout the series.
- Important concepts from **Number**, **Algebra**, **Measurement** and **Space** are explored.
- **Measurement** concepts and activities are introduced and investigated.
- **Statistics and Probability** concepts are presented for revision and extension.

- A **tables** program for each of addition, subtraction and multiplication is included. It is important for students to learn addition and multiplication tables by heart.

3 Contents

Activities at the bottom of each page

Unit	Content	Activity type
1:1/2 1:3/4	+ 2 Personal measures	+ tables Measure
2:1/2 2:3/4	+ 3, + 5 + 3, + 4	+ tables + tables
3:1/2 3:3/4	+ 4, + 6 + 6, + 7	+ tables + tables
4:1/2 4:3/4	Chance + 7, + 9	Chance + tables
5:1/2 5:3/4	+ 8, + 9 Chance	× tables Chance
6:1/2 6:3/4	Language Linking + and −	ID Card B Concept
7:1/2 7:3/4	× 2, × 0 × 10	× tables × tables
8:1/2 8:3/4	× 5, × 2 × 5, × 10	× tables × tables
9:1/2 9:3/4	× 1, × 10 Subtracting 9	× tables Concept
10:1/2 10:3/4	Linking + and − Skip counting	Concept × tables
11:1/2 11:3/4	Rounding (nearest 10) Problem solving	Concept Strategy time
12:1/2 12:3/4	Rounding (nearest 100) Writing fractions	Concept Concept
13:1/2 13:3/4	Using number lines Addition grid	− tables + tables
14:1/2 14:3/4	× 5, × 10 Perimeter	× tables Measure
15:1/2 15:3/4	Combinations to 10 and 13 + 10, + 10	Concept + tables
16:1/2 16:3/4	Time Time	Measure Concept
17:1/2 17:3/4	× 3 Skip counting	× tables × tables
18:1/2 18:3/4	Chance × 2, × 4	Chance × tables
19:1/2 19:3/4	Chance × 3, × 4	Chance × tables

Unit	Content	Activity type
20:1/2 20:3/4	Language Chance	ID Card B Chance
21:1/2 21:3/4	× 3, × 4 7 −, 8 −	× tables − tables
22:1/2 22:3/4	11 −, 12 − 13 −, 14 −	− tables − tables
23:1/2 23:3/4	Time + 3, − 3	Measure + / − tables
24:1/2 24:3/4	rows of, groups of − 10, − 10	Concept − tables
25:1/2 25:3/4	Linking × and ÷ + 4, + 3	Concept + tables
26:1/2 26:3/4	Language + 5, + 5	ID Card A + tables
27:1/2 27:3/4	Language Linking × and ÷	ID Card A Concept
28:1/2 28:3/4	Language ÷ 2, ÷ 4	ID Card B ÷ tables
29:1/2 29:3/4	+ 3, + 4 + 5, + 6	+ tables + tables
30:1/2 30:3/4	Chance Trading with blocks, +	Chance Concept
31:1/2 31:3/4	13 −, 14 − Chance	− tables Chance
32:1/2 32:3/4	Place value × 10, × 5, × 2	Concept × tables
33:1/2 33:3/4	× 3, × 4, × 5 24 +, 32 +	× tables + tables
34:1/2 34:3/4	Language Place value	ID Card B Concept
35:1/2 35:3/4	Language Language	ID Card A ID Card B
36:1/2 36:3/4	15 −, 16 − − 2, − 4	− tables − tables
37:1/2 37:3/4	+ 2, + 3, + 4 Personal measures	+ tables Measure
Answers	These can be found in the middle of this book on pages A1 to A16.	

Arithmetic card

	A	B	C	D	E	F	G	H	I	J
1	11	3	7	20	4	2	7	50	37	$6
2	17	7	4	16	25	10	11	80	93	$1
3	14	1	10	12	64	14	3	20	16	$7
4	19	5	3	17	16	4	15	70	55	$3
5	16	8	5	13	81	12	19	40	100	$8
6	12	6	8	18	1	16	1	60	71	$5
7	18	10	2	11	36	8	9	100	48	$9
8	15	4	6	14	9	20	17	30	82	$2
9	20	9	1	19	100	18	5	90	64	$10
10	13	2	9	15	49	6	13	10	29	$4

How to use this card

If students were told to 'subtract B from A', they would write:

1. 11 − 3 = 8
2. 17 − 7 = 10
3. 14 − 1 = 13
4. 19 − 5 = 14
5. 16 − 8 = 8
6. 12 − 6 = 6
7. 18 − 10 = 8
8. 15 − 4 = 11
9. 20 − 9 = 11
10. 13 − 2 = 11

Other instructions might be:

- **Multiply column B by 5.**
- **Subtract column F from column I.**
- **Halve column F.**
- **What multiplied by 10 gives column H?**
- **What is left from $30 if I spend the amount in column J?**
- **Double column C.**
- **Add column B and C.**
- **Subtract 10 from column I.**
- **Add 10 to column E.**
- **What must be added to column D to make 20?**

The applications of this card are endless.

ID card A

Do not write on this card.

1 **m** stands for m ______	2 **cm** stands for c ______	3 **mm** stands for m ______	4 **L** stands for l ______	5 **kg** stands for k ______
6 **min** stands for m ______	7 1 cm, 1 cm 1 square c ______	8 1 cm, 1 cm, 1 cm 1 cubic c ______	9 **2, 4, 6, …** are the e ______ numbers	10 **1, 3, 5, …** are the o ______ numbers
11 **1st, 2nd, …** are the o ______ numbers	12 **816** has 3 d ______	13 **×** stands for t ______	14 **6 groups of 2** **or** **6 rows of 2** means 6 ______ 2	15 **÷** stands for d ______
16 **How many groups of …?** or **share between** means d ______	17  means is e ______ to	18 0 1 2 3 4 n ______ l ______	19 **7 + 9 = 16** is a n ______ s ______	20 7 tens 2 ones is a n ______ expander
21 Th H T U a ______	22 Animals: birds 7, monkeys 5, lions 4 t ______	23 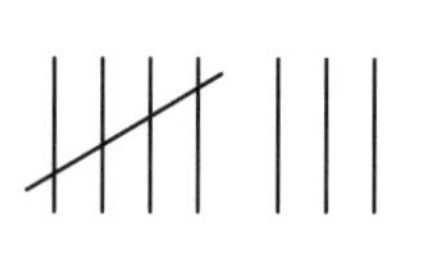 Balls Lost: May, June, July stands for one ball p ______ graph	24 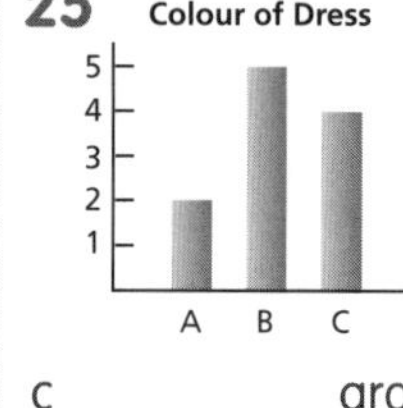 t ______	25 Colour of Dress 5, 4, 3, 2, 1 A B C c ______ graph
26 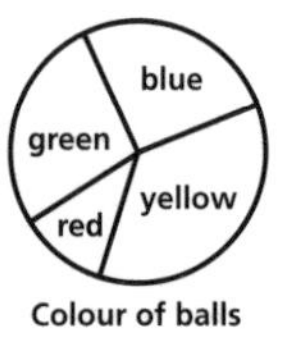 blue, green, yellow, red Colour of balls s ______ graph	27 5:30 am d ______ watch	28 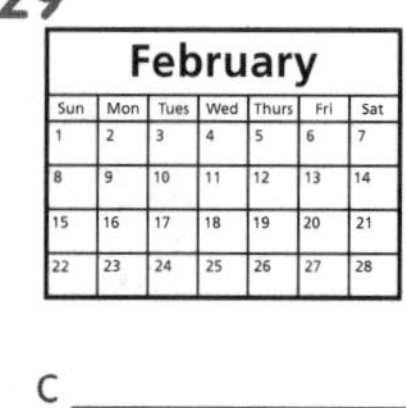 a ______ clock	29 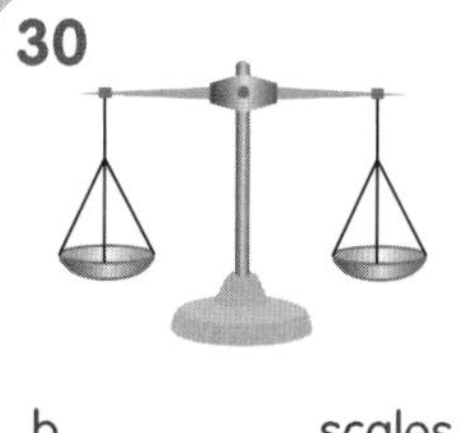 February c ______	30 b ______ scales

See page A1 for answers.

 ISBN 978 0 6557 0883 4

ID card B

Do not write on this card.

1 a ________

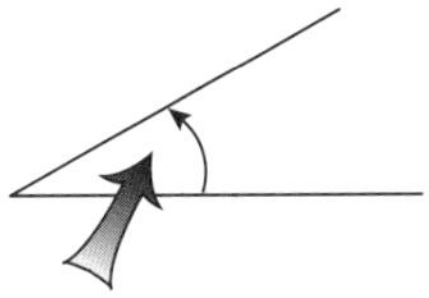

2 r ________ angle

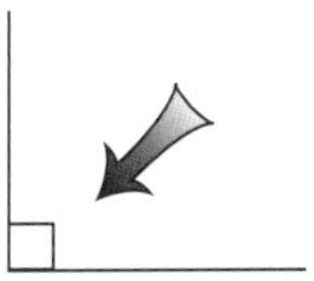

3 p ________ lines

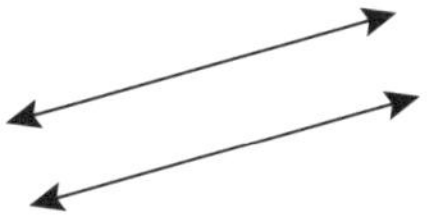

4 p ________ lines

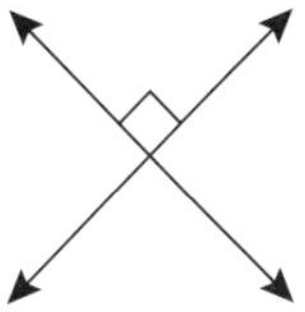

5 t ________

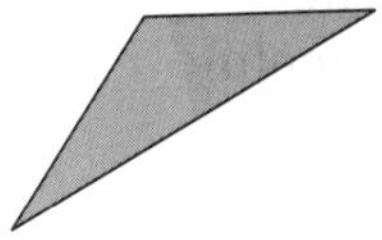

6 s ________

7 r ________

8 r ________

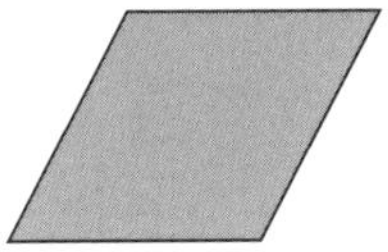

9 t ________

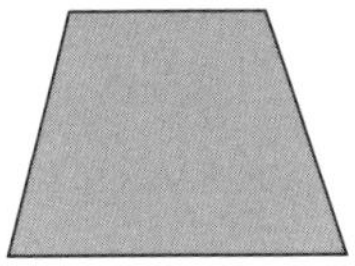

10 p ________

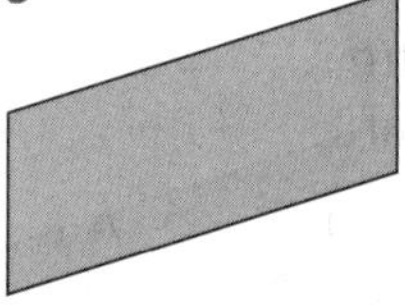

11 q ________

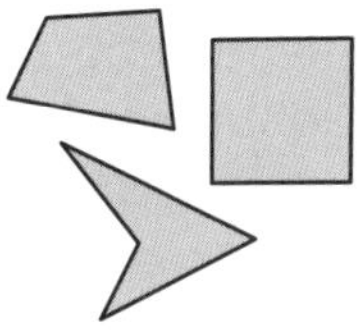

12 p ________

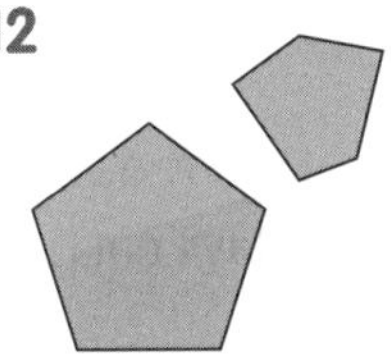

13 h ________

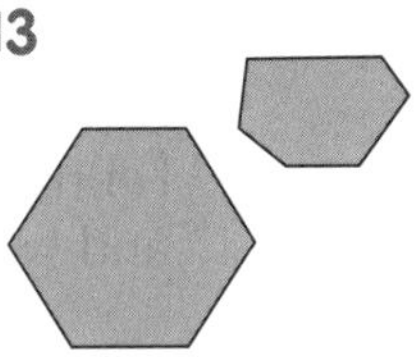

14 c ________

15 o ________

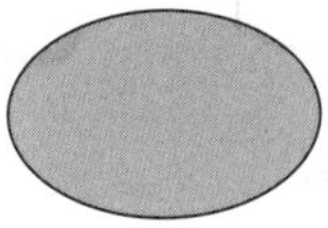

16 r ________ shapes

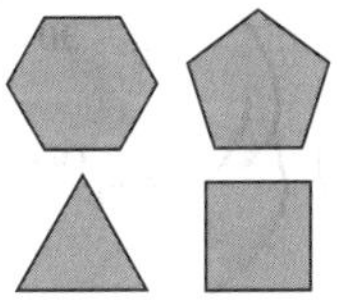

17 i ________ shapes

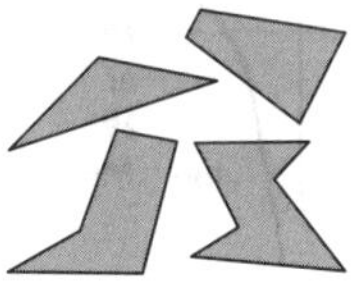

18 s ________ c ________

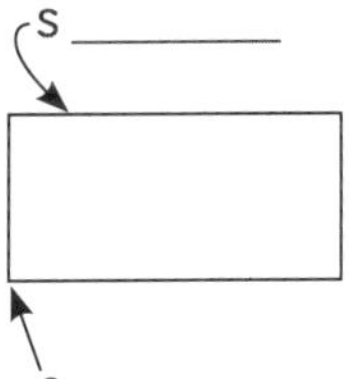

19 line of s ________

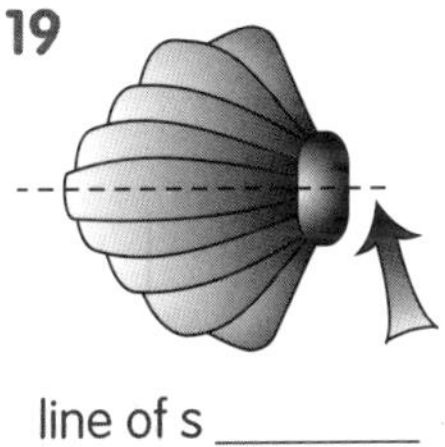

20 net of a c ________

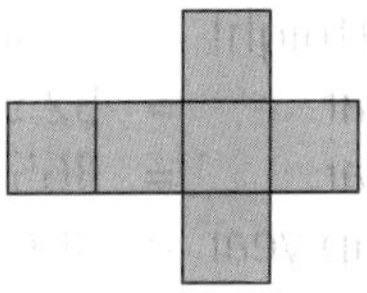

21 c ________

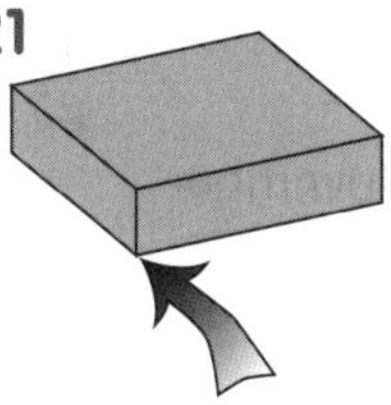

22 e ________

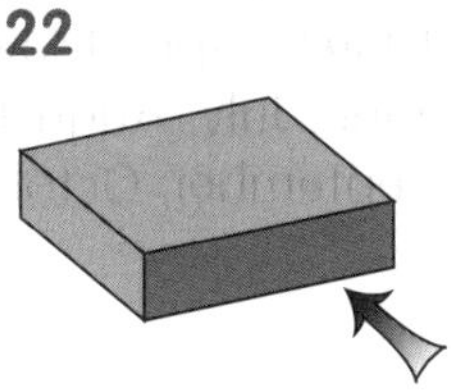

23 f ________

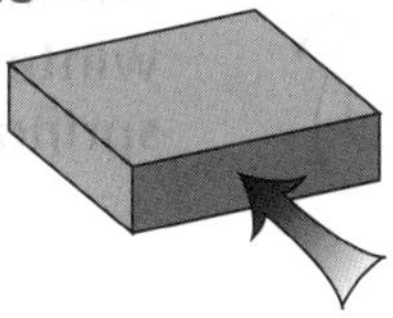

24 c ________

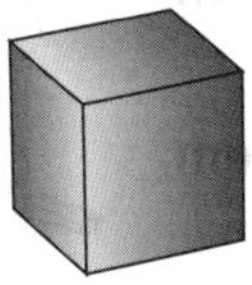

25 p ________

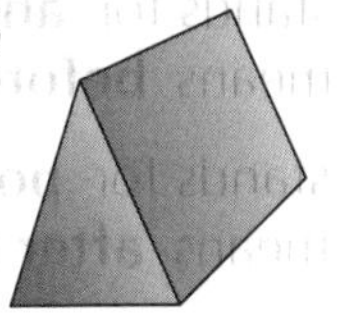

26 p ________

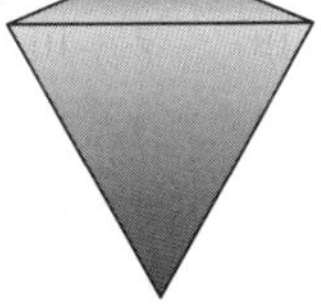

27 b ________

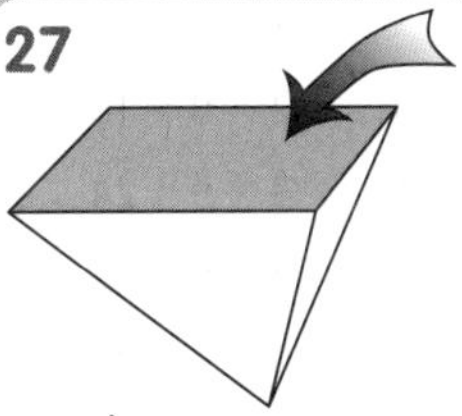

28 c ________

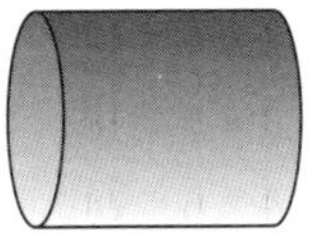

29 c ________

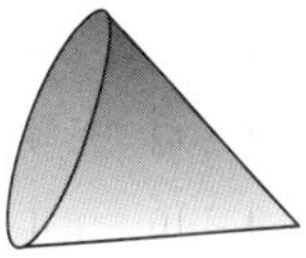

30 s ________

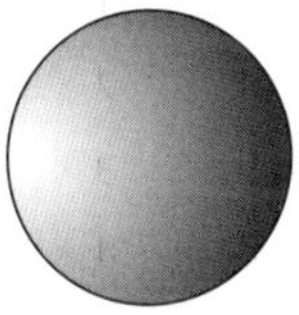

See page A1 for answers.

Tables of number and measurement

Length

1 centimetre = 10 millimetres
1 metre = 100 centimetres
1 metre = 1000 mm
1 kilometre = 1000 metres

Mass

1 kilogram = 1000 grams

Capacity and volume

1 litre = 1000 millilitres

1 litre of water has a mass of 1 kg.
A carton of milk holds 1 L.

A teaspoon holds about 5 mL.

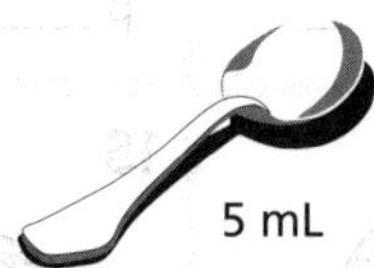

Time

1 minute = 60 seconds
1 hour = 60 minutes
1 day = 24 hours
1 week = 7 days
1 fortnight = 2 weeks
1 year = 52 weeks
1 year = 365 days
1 leap year = 366 days
1 decade = 10 years
1 century = 100 years

am stands for **ante meridiem**.
am means **before midday**.

pm stands for **post meridiem**.
pm means **after midday**.

You must learn your tables!

The **freezing point** of water is **0°C**.
The **boiling point** of water is **100°C**.

A temperature of **5°C** is **a cold day**.
A temperature of **35°C** is **a hot day**.

Roman numerals

1	= I	**6**	= VI	**20**	= XX	**90**	= XC
2	= II	**7**	= VII	**30**	= XXX	**100**	= C
3	= III	**8**	= VIII	**40**	= XL	**200**	= CC
4	= IV	**9**	= IX	**50**	= L	**500**	= D
5	= V	**10**	= X	**60**	= LX	**1000**	= M

Months of the year

Thirty days has September, April, June and November. All the rest have thirty-one, except February alone, which has twenty-eight days clear and twenty-nine days each leap year.

You can use the knuckles of your hands to find the number of days in each month.

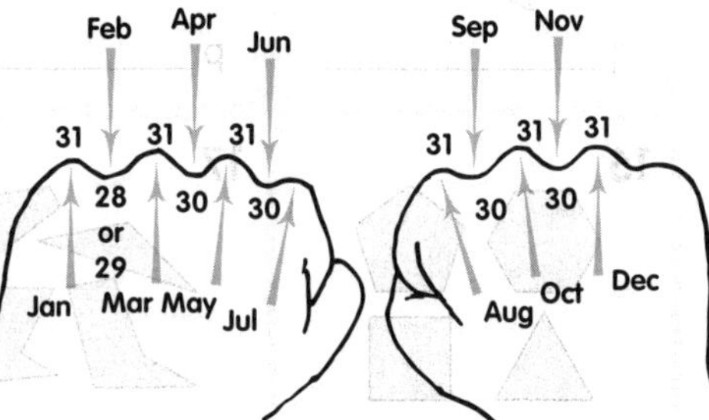

Seasons

summer: December, January, February
autumn: March, April, May
winter: June, July, August
spring: September, October, November

Multiplication tables

1 × 2 = 2	1 × 3 = 3	1 × 4 = 4	1 × 5 = 5	1 × 6 = 6	1 × 7 = 7	1 × 8 = 8	1 × 9 = 9	1 × 10 = 10
2 × 2 = 4	2 × 3 = 6	2 × 4 = 8	2 × 5 = 10	2 × 6 = 12	2 × 7 = 14	2 × 8 = 16	2 × 9 = 18	2 × 10 = 20
3 × 2 = 6	3 × 3 = 9	3 × 4 = 12	3 × 5 = 15	3 × 6 = 18	3 × 7 = 21	3 × 8 = 24	3 × 9 = 27	3 × 10 = 30
4 × 2 = 8	4 × 3 = 12	4 × 4 = 16	4 × 5 = 20	4 × 6 = 24	4 × 7 = 28	4 × 8 = 32	4 × 9 = 36	4 × 10 = 40
5 × 2 = 10	5 × 3 = 15	5 × 4 = 20	5 × 5 = 25	5 × 6 = 30	5 × 7 = 35	5 × 8 = 40	5 × 9 = 45	5 × 10 = 50
6 × 2 = 12	6 × 3 = 18	6 × 4 = 24	6 × 5 = 30	6 × 6 = 36	6 × 7 = 42	6 × 8 = 48	6 × 9 = 54	6 × 10 = 60
7 × 2 = 14	7 × 3 = 21	7 × 4 = 28	7 × 5 = 35	7 × 6 = 42	7 × 7 = 49	7 × 8 = 56	7 × 9 = 63	7 × 10 = 70
8 × 2 = 16	8 × 3 = 24	8 × 4 = 32	8 × 5 = 40	8 × 6 = 48	8 × 7 = 56	8 × 8 = 64	8 × 9 = 72	8 × 10 = 80
9 × 2 = 18	9 × 3 = 27	9 × 4 = 36	9 × 5 = 45	9 × 6 = 54	9 × 7 = 63	9 × 8 = 72	9 × 9 = 81	9 × 10 = 90
10 × 2 = 20	10 × 3 = 30	10 × 4 = 40	10 × 5 = 50	10 × 6 = 60	10 × 7 = 70	10 × 8 = 80	10 × 9 = 90	10 × 10 = 100

Examples of measurements

1

2

3

4

5

6

1
- The width of your finger is about 1 cm.
- The length of a place-value tens block is 10 cm.

2
- The height of the girl is a little more than 1 m.

3
- The container of milk holds 2 L.
- The can of Fizz holds 375 mL.
- The teaspoon holds 5 mL.

4
- The boy has a mass of 40 kg.
- The margarine has a mass of 500 g.

5
- The area of the newspaper is about the same as the table.

6
- 30 degrees is a hot day.
- 3 degrees is a very cold day.

Use the pictures above to estimate the answers to these questions.

1 **a** How high is the glass?
b How wide is the table?

2 **a** How wide is the clothes line?
b How tall is the woman?

3 **a** How much will the bucket hold?
b How much will the cup hold?

4 **a** What is the mass of the dog?
b What is the mass of 2 L of milk?

5 **a** How many sheets of newspaper would cover the area of the window?
b How many place-value ones would cover the area of the top of the matchbox?

6 **a** What is the temperature on a very hot day?
b What is the temperature on a cool day?

 • *AUSTRALIAN SIGNPOST MATHS 3 MENTALS* • ISBN 978 0 6557 0883 4

1:1 [ ] out of 9

1. In **a** to **f**, write the number modelled.

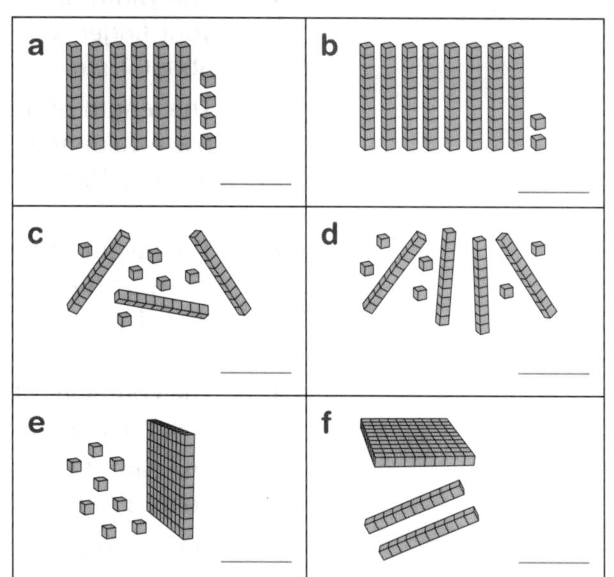

2. How many place-value ones blocks make one tens block? ______
3. The sign = means "is e ______ to".
4. Complete the pattern.

   2, 4, ______, ______, 10, 12
5. The tally 𝍸 || stands for: ______
6. 

   This is a n______________ l______________.
7. How many digits in 13? ______
8. Josh had 5 toy trucks and was given 4 more by his Pa. How many has he now? ______
9. The numeral for twelve. ______

1:2 [ ] out of 15

1. 2 + 8 ______
2. 17 + 3 ______
3. 20 − 5 ______
4. 20 − 7 ______
5. $\begin{array}{r} 23 \\ +\ 3 \\ \hline \end{array}$
6. Digits in 184. ______
7. \$7 + \$7 ______
8. 7, 9, 11, ______, ______
9. Take 5 from 15. ______
10. $\begin{array}{r} 19 \\ -\ 4 \\ \hline \end{array}$
11. 

    The number sentence is:

    8 − ______ = ______
12. Colour one half of this rectangle.

13. What time is shown?

    a 

    ______ past ______

    b 

    ______ past ______
14. Circle half of this group. 
15. How many in each box if we share the balls fairly? ______

    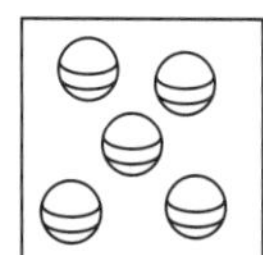 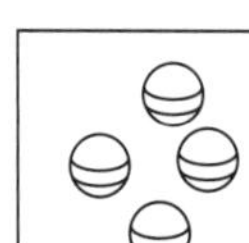 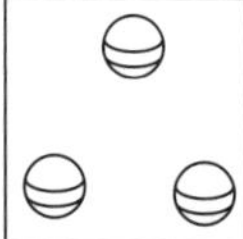 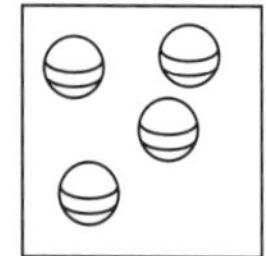

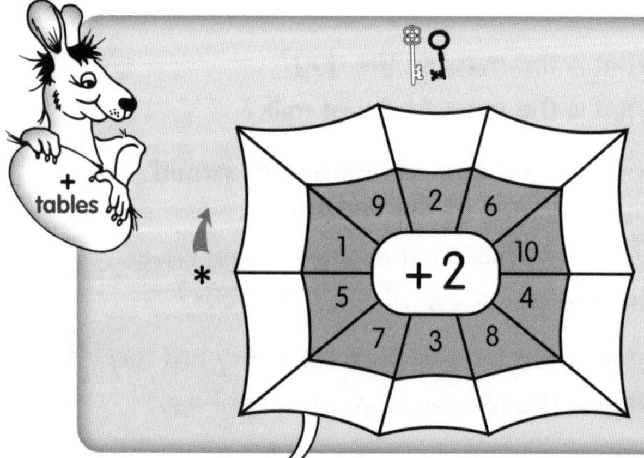

- Odd numbers end in 1, 3, 5, 7 or 9.
- Even numbers end in 2, 4, 6, 8 and 0.
- If you add an even number to an even number the answer is always even.

What happens when you add these types of numbers?

odd + odd = ______________

even + odd = ______________

1:3 ☐ out of 19

1. 24 − 10 ____
2. 13 + 4 ____
3. 25 − 6 ____
4. 27 + 6 ____
5. 18 + 4 ____
6. 2, 4, 6, ____, ____
7. Add 19 and 7. ____
8. 23 minus 4. ____
9. Write 19 in words. ____
10. 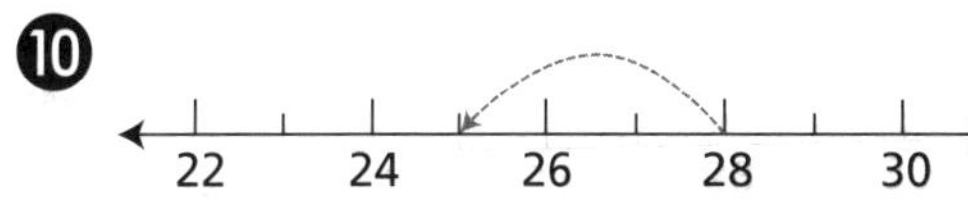

 a 28 – 3 ____ b 28 – 5 ____
11. 11 more than 17. ____
12. The numeral for fifty-two. ____
13. What month comes after December? ____
14. Use the split strategy to find:

 a 13 + 12 = ____ b 23 + 26 = ____

 c 45 + 51 = ____ d 33 + 42 = ____
15. Four equal q________ make 1 whole. Colour one quarter. 
16. Wednesday is the ____ th day of the week.
17. Complete this pattern.

 42, 44, 46, ____, ____
18. Circle one quarter of this group. 
19. The value of these coins.    

 ____

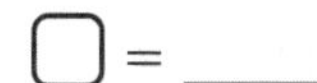

1:4 ☐ out of 10

1. 10 + ☐ = 16 ☐ = ____
2. 72, 62, 52, ____, ____, ____
3. 3, 6, 9, ____, ____, ____, ____
4. Use the split strategy. Show the working.

 15 + 38 = (10 + 30) + (5 + ____)

 = ____ + ____

 = ____
5. The difference between 20 and 17 = ____
6. If 8 + 12 = 20 then 12 + 8 = ____

8	12
20	

 20 − 8 = ____

 20 − 12 = ____
7. Three different counting numbers that add to give 7. ____
8. How many days altogether in summer? ____
9. Which season has the least days?

 ____
10. Two weeks is also called a fortnight.

 Weeks in 4 fortnights? ____

Challenge

Write as many different number patterns as you can fit in this space.

Measure

Fill out this table about yourself, a relative or a friend.

Name: ____ **Date:** ____

Age: ____	Mass: ____ kg	Shoe size: ____
Height: ____ cm	Waist: ____ cm	Neck size: ____ cm

2:1 [ ] out of 7

1 In **a** to **f**, write the number modelled.

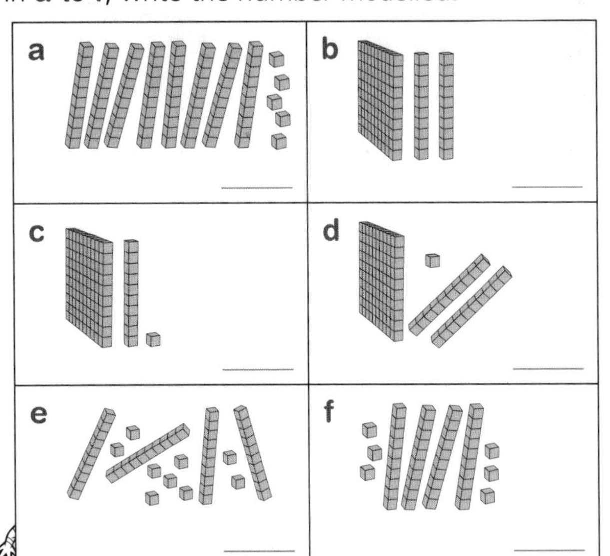

2 **a** 16 + 4 = ______ **b** 15 + 5 = ______

3 How many ones in 3 tens? ______

4 The numeral for eighteen. ______

5 Days in 1 week. ______

6

6 rabbits are shared by 2 families.

6 ÷ 2 = ______ Each has ______.

7 How many minutes in half an hour? ______

2:2 [ ] out of 16

1 12 + 8 ______

2 20 − 6 ______

3 17 + ______ = 20

4 38 + ______ = 40

5
```
  12
+  5
```

6 4 × 2 ______

7 5 × 2 ______

8 10 shared by 2. ______

9 8 shared by 2. ______

10
```
  20c
−  5c
```

11 Colour one quarter.

12 Use the split strategy to find:

a 31 + 52 = ______ **b** 54 + 24 = ______

c 83 + 14 = ______ **d** 33 + 35 = ______

13 Use the jump strategy to find:

a 39 + 17 = ______

39

b 68 + 24 = ______

68

14 How many fingers on 3 hands? 3 × 5 = ______

15 A B C

A is a ______.

B is a ______.

C is a ______.

16 Draw 4 groups of 3 balls. Total balls = ______

+ tables

*

6	3	1	10	7	4	9	8	2	5

+3

*

6	3	1	10	7	4	9	8	2	5

+5

odd + odd = ____

 • *AUSTRALIAN SIGNPOST MATHS 3 MENTALS* • ISBN 978 0 6557 0883 4

2:3 ☐ out of 13

1. 40 + 6 ____
2. 37 + 3 ____
3. 12 + 14 ____
4. 16 + 31 ____
5. Double 13. ____
6. Halve 22. ____
7. 56 + 10 ____
8. 28 + 12 ____
9. **a** 28 + 25 = ____ (number line starting at 28)

   b 39 + 34 = ____ (number line starting at 39)

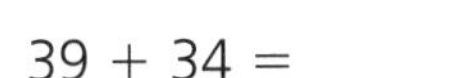

10.      

    a 2 groups of 4 = ____ apples

    b 3 groups of ____ = ____ apples

11. **a** 0 1 2 3 4 5 6 7 8 9 10

    a The difference between 8 and 5 ____

    b The difference between 28 and 25 ____

12. Write two linking number sentences for

    4 + 9 = ____

☐ − ☐ = ☐

☐ − ☐ = ☐

13. Circle the **heavier** object.

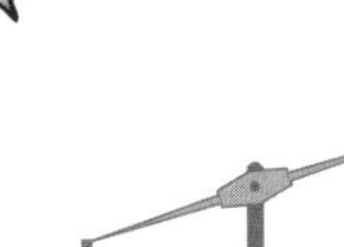
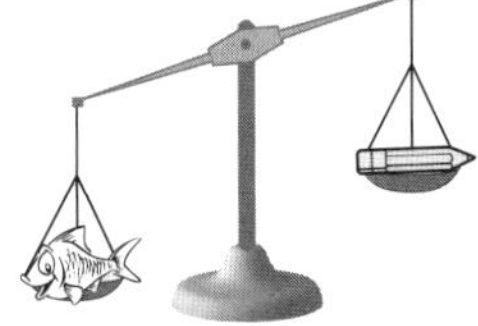

2:4 Extension ☐ out of 6

1. **a** 7 + ☐ = 15 ☐ = ____

   b ☐ + 16 = 24 ☐ = ____

2. In a race, I am 4th out of 5. How many are:

   a in front of me? ____

   b behind me? ____

3. Use the code to work out this message.

A	B	C	D	E	I	K	N	S
1	2	3	4	5	6	7	8	9

11 − 9	2 + 3

3 + 4	10 − 4	8 + 0	12 − 8

4. What is the time half an hour after quarter past three? ____

5. Colour one quarter of this shape.

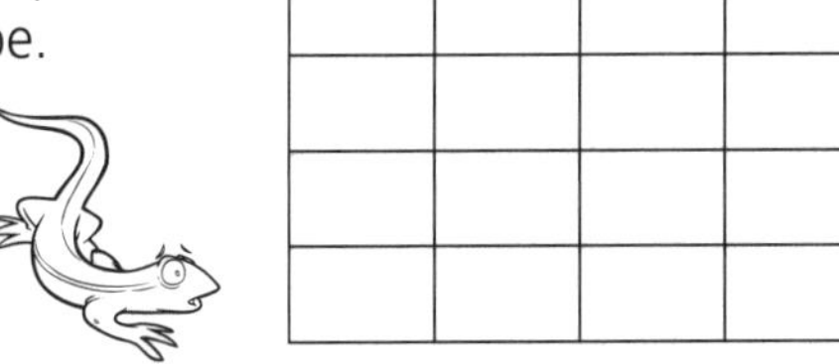

6. 2 + 8 + 3 + 7 + 1 + 9 ____

Challenge

Write different number sentences that equal 15.

____

____

____

+ tables

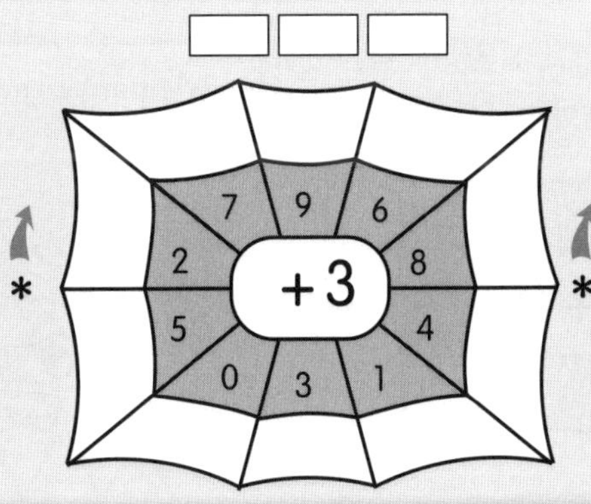

Memorise these facts.

3:1 out of 7

1 a 

b

_____ past _____ _____ to _____

2 a 3, 6, 9, ____, ____, ____, ____

b 4, 8, 12, ____, ____, ____, ____

c 7, 14, 21, ____, ____, ____, ____

3 a 4 shared by 2 = ____ so 4 ÷ 2 = ____

b 6 shared by 2 = ____ so 6 ÷ 2 = ____

4 Summer had $20 and spent $10 to buy a toy bear. How much does she have left? ____

5 a 1 + 9 = ____ b 9 + 1 = ____

c 10 − 1 = ____ d 10 − 9 = ____

6 What is:

a behind the book? ____

b in front of the jug? ____

7 I had 10 toy dinosaurs. I sold 6. How many were left? ____

3:2 out of 17

1 14 + 6 ____

2 20 − 3 ____

3 15 + ____ = 20

4 48 + ____ = 50

5
$$\begin{array}{r} 19 \\ +\ 8 \\ \hline \end{array}$$

6 3 × 2 ____

7 6 × 2 ____

8 4 shared by 2. ____

9 12 shared by 2. ____

10
$$\begin{array}{r} 30\text{c} \\ -\ 4\text{c} \\ \hline \end{array}$$

11 Count on to the next 10 first, to find:

a 6 + ____ = 12 b 19 + ____ = 22

c 5 + ____ = 14 d 17 + ____ = 26

12 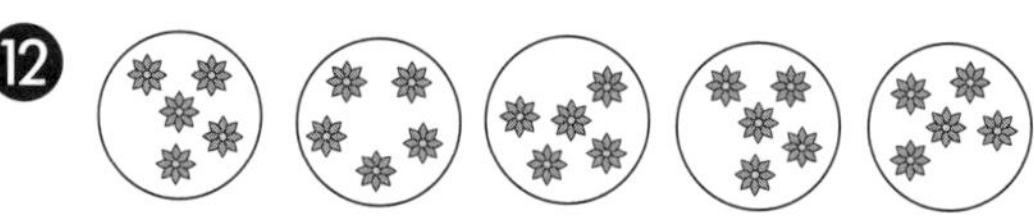

____ groups of ____ = ____

13 a 26 + 9 − 9 = ____ b 39 + 5 − 5 = ____

14 Complete these near doubles.

3 + 4 = ____ 7 + 8 = ____ 9 + 8 = ____

15 Use bridging to ten to find:

9 + 5 = ____ 8 + 6 = ____ 9 + 3 = ____

16 How many groups of 10 are there in:

a 46? ____ b 69? ____ c 37? ____

17 Circle the largest number. Underline the smallest number.

a 503, 530, 550, 500

b 460, 406, 462, 426

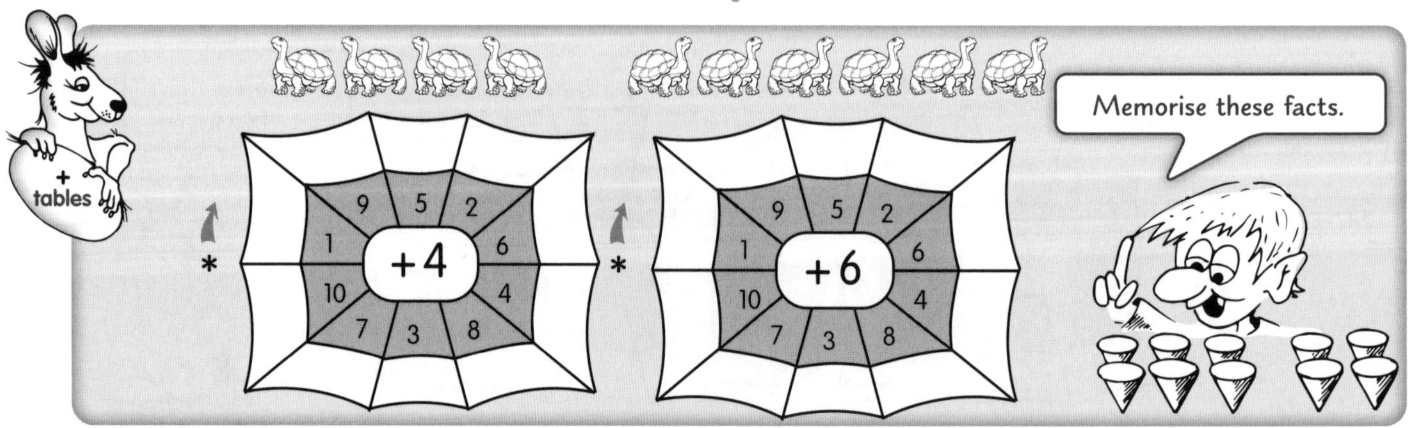

ISBN 978 0 6557 0883 4

3:3 ☐ out of 10

1
$$\begin{array}{r} 4 \\ 6 \\ +\ 2 \\ \hline \end{array}$$

2
$$\begin{array}{r} 6 \\ 3 \\ +\ 7 \\ \hline \end{array}$$

3
$$\begin{array}{r} 8 \\ 4 \\ +\ 2 \\ \hline \end{array}$$

4 Bridge to ten to find:

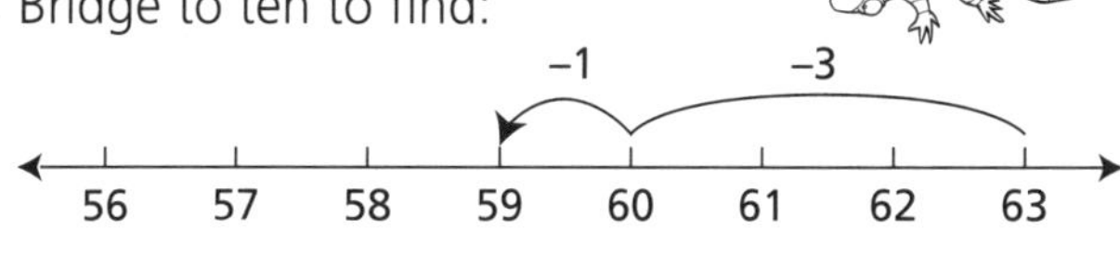

a 63 − 4 = ________ **b** 63 − 7 = ________

c 61 − 3 = ________ **d** 61 − 5 = ________

5 Yoo-Jin had 5 pairs of shoes. How many shoes does he have altogether? ________

6 _____ + 18 = 21 _____ + 15 = 23

7 How many groups of 100 are in:

a 325? _____ **b** 729? _____

8 Count on to the next 10 first, to find:

a 9 + _______ = 15 **b** 18 + _______ = 26

c 16 + _______ = 21 **d** 15 + _______ = 23

9 Write the total for each.

a

5	5
10	

b

5	5	5	5	5

c

4	4	4	4

d

4	4	4

10 **a** How many days in 1 week? ______

b How many months in 1 year? ______

3:4 Extension ☐ out of 5

1 Would a 4-digit number be bigger than a 3-digit number? ________

2 **a** ☐ − 13 = 4 ☐ = _____

b 20 − ☐ = 6 ☐ = _____

3

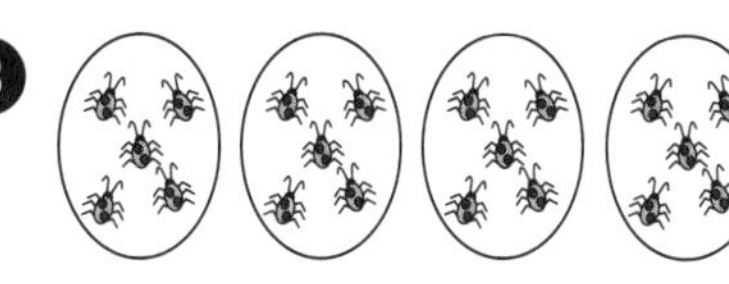

a How many bugs in 4 groups? ________

b How many bugs in 5 groups? ________

c How many bugs in 8 groups? ________

4 **a** The smallest area below. ________

b The largest area below. ________

A B C

5 Minutes in 3 hours? ________

Challenge

Show how you can make 50c in different ways.

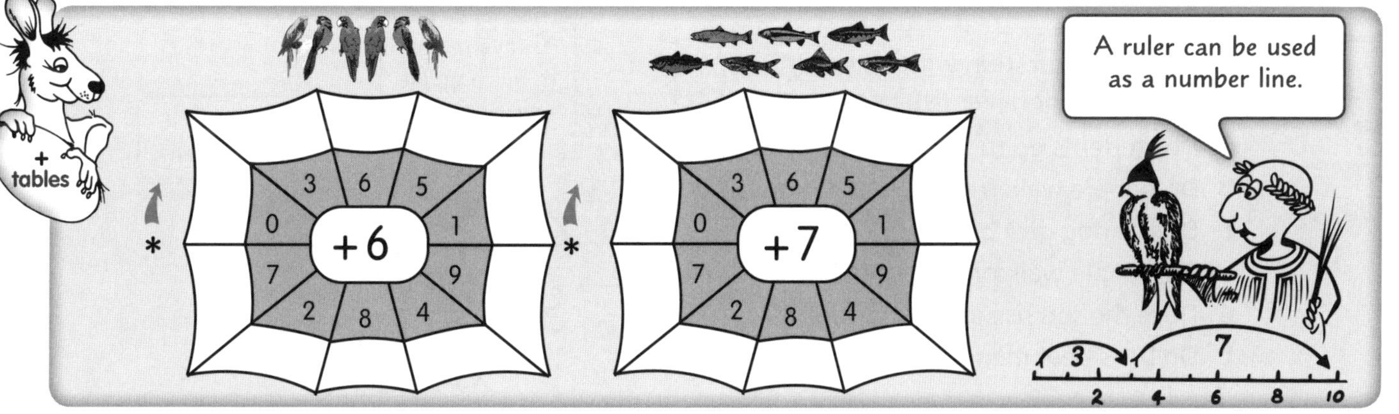

4:1 ☐ out of 15

1. 13 + 7 = _____
2. 20 − 9 = _____
3. 18 + _____ = 20
4. 28 + _____ = 30
5. $\begin{array}{r} 19 \\ +\ \ 6 \\ \hline \end{array}$
6. 8 × 2 _____
7. 10 × 2 _____
8. 20 shared by 2. _____
9. 16 shared by 2. _____
10. $\begin{array}{r} 30\text{c} \\ -\ \ 6\text{c} \\ \hline \end{array}$

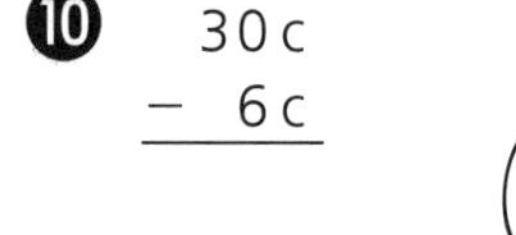

11. Count on to the next 10 first, to find:
    - **a** 9 + _______ = 11 **b** 19 + _______ = 21
    - **c** 9 + _______ = 13 **d** 19 + _______ = 23
12. What is to the:

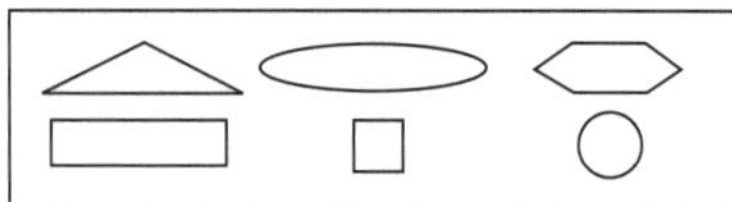

    - **a** left of the hexagon? _______
    - **b** right of the rectangle? _______
13. Is it possible to grow a tail and long ears if I eat carrots? _______
14.
    - **a** 2, 4, 6, _____, _____, _____, _____
    - **b** 100, 90, 80, _____, _____, _____, _____
    - **c** 102, 104, 106, _____, _____, _____
    - **d** 1000, 900, 800, _____, _____, _____
15. I have 6 pairs of socks. Draw them below.

    6 groups of 2 = _____ 6 × 2 = _____

4:2 ☐ out of 18

1. 27 − 5 = _____
2. 26 + 4 = _____
3. 12 × 2
4. Halve 14. _____
5. $\begin{array}{r} 27 \\ +\ \ 5 \\ \hline \end{array}$
6. 9 × 2 _____
7. 14 × 2 _____
8. 12 shared by 3. _____
9. 16 shared by 4. _____
10. $\begin{array}{r} 40\text{c} \\ -\ \ 9\text{c} \\ \hline \end{array}$
11. Which is large: 108 or 180? _______
12. Half of 16. _______
13. How many edges has this shape? _______

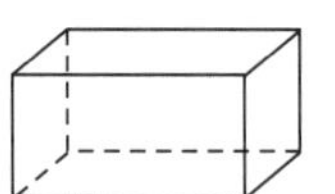

14.
    - **a** 6, 12, 18, _____, _____, _____, _____
    - **b** 10, 20, 30, _____, _____, _____, _____
    - **c** 9, 18, 27, _____, _____, _____, _____
15. Luke is 9. How old will he be in:
    - **a** 9 years? _______ **b** 10 years? _______
    - **c** 16 years? _______ **d** 31 years? _______
16. Is there an even chance of spinning a **B**? _______

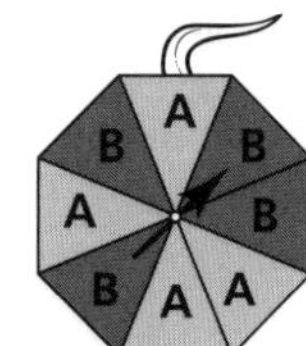

17. Jasmine has 4 plates, 4 cups and 4 spoons. How many altogether? _______
18. How many fingers on:
    - **a** 6 students? _______
    - **b** 10 students? _______
    - **c** 8 students? _______

Answer each question with always (A), sometimes (S) or never (N).

- **a** Do students go to school on Saturday? _______
- **b** Do people walk from here to China? _______
- **c** Does a dog give birth to a cat? _______
- **d** Do people walk on the moon? _______
- **e** Does the sun rise in the morning? _______
- **f** Do boys beat men at running? _______

Always means it will happen every time.

Never means it cannot happen.

 • *AUSTRALIAN SIGNPOST MATHS 3 MENTALS* • ISBN 978 0 6557 0883 4

4:3 out of 12

1
```
  8
  3
  2
+ 7
```

2
```
  5
  1
  9
+ 5
```

3
```
  8
  0
  9
+ 2
```

4 90, 80, 70, ______, ______, ______

5 6 hundreds 5 tens 3 ones = ______

6 Describe this object.

______

______

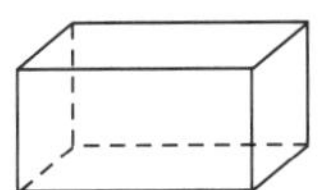

7 a 93, 83, 73, ______, ______, ______, ______

b 695, 690, 685, ______, ______, ______

c 529, 539, 549, ______, ______, ______

8 When we count by 2s from zero, the numbers end in ______.

9 Certain, impossible, likely or unlikely?

"The next person I meet will be from Mars."

______

10 Use the split strategy to find:

a 16 + 13 = ______ b 41 + 53 = ______

c 52 + 46 = ______ d 58 + 21 = ______

11 Use the jump strategy to find:

a 29 + 8 = ______ b 48 + 23 = ______

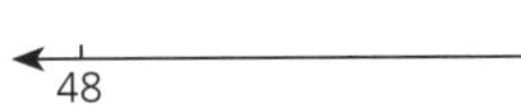

12 a 56 + ______ = 60 b 71 + ______ = 80

4:4 Extension out of 8

1 200, 250, 300, ______, ______, ______, ______

2 7 + 13 + 12 + 8 + 1 + 9 ______

3 28 − 4 − 4 − 4 − 4 − 4 ______

4 Weeks in 1 year. ______

5 Complete the pattern.

15, 12, 9, ______, ______

6 Can you work out the message?

Code

A	E	I	L	M	S	T
@	?	/	$	^	•	*

•	^	/	$	?	@	*	^	?

7 Complete:

a 20 + ______ = 100

b 100 − 20 = ______

8 Days in 1 fortnight? ______

Challenge

23 − 6 = ______

Explain how you found your answer.

______

______

______

______

______

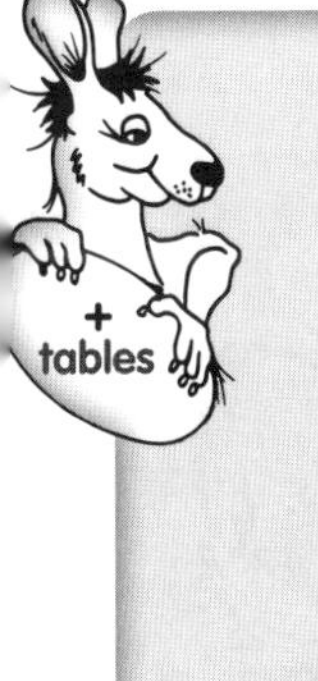

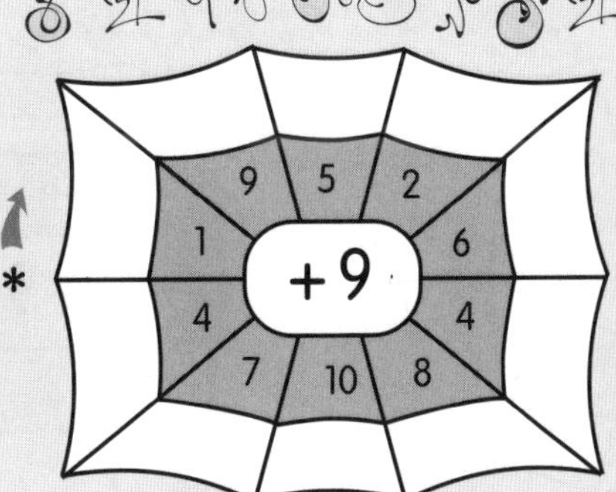

5:1 out of 15

1. 28 + 2 = ____
2. 20 − 5 = ____
3. 11 + ____ = 20
4. 25 + ____ = 30
5. 22 + 24
6. 3 × 10 ____
7. Half of 12. ____
8. 12 ÷ 2 ____
9. 4 ÷ 2 ____
10. 38c − 14c ____

11. *Impossible*, *unlikely*, *likely*, *certain*

    Which of these describes the chance of throwing a 6 on a die? ____

12. Write the answer and fill in the numeral expander.

    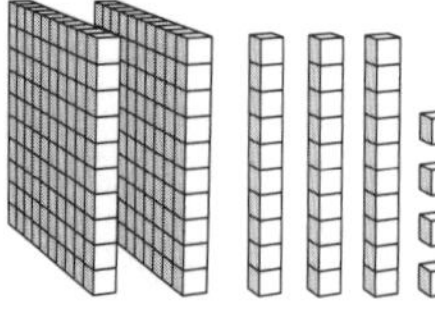

    = ____

	hundreds		tens		ones

13. Write the number above in words.

    ____

    ____

14. How many faces (flat surfaces) has:

    a A? ____

    b B? ____

    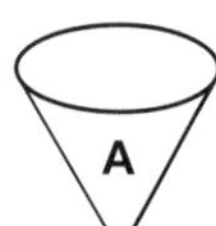

    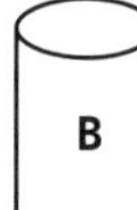

15. Show how you jump to the next ten to find:

    a 7 + 5 = ____

5:2 out of 18

1. 23 + 20 = ____
2. 40 − 6 = ____
3. 34 + ____ = 40
4. 34 + ____ = 41
5. 62 + 12
6. 8 × 2 ____
7. Half of 18. ____
8. 16 ÷ 2 ____
9. 14 ÷ 2 ____
10. 38c − 23c ____

11. Is there an even chance of landing on **A**?

    ____

12. Order from smallest to largest.

    391, 902, 535 ____, ____, ____

13. Write six hundred and four as a numeral. ____

14. Write the numeral for:

    ____

    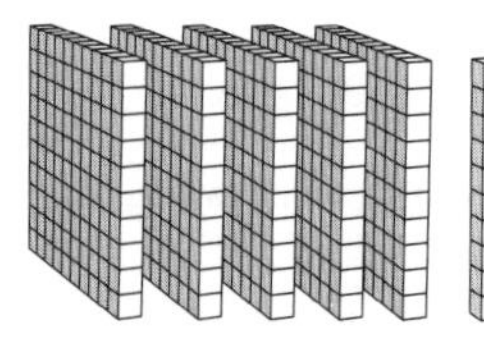

15. The season after summer? ____

16. **a** 3, 6, 9, ____, ____, ____, ____

    b 76, 66, 56, ____, ____, ____, ____

    c 172, 162, 152, ____, ____

    d 430, 435, 440, ____, ____

17. Draw the line of symmetry.

18. November is the ____ month of the year.

+ tables

0	8	4	6	1	9	7	3	10	5

+8

0	8	4	6	1	9	7	3	10	5

+9

2 + 9 is the same as 9 + 2.

 • *AUSTRALIAN SIGNPOST MATHS 3 MENTALS* • ISBN 978 0 6557 0883 4

5:3 ☐ out of 9

1 $\begin{array}{r} 35 \\ +\ 6 \\ \hline \end{array}$ **2** $\begin{array}{r} 26 \\ +\ 8 \\ \hline \end{array}$ **3** $\begin{array}{r} 44 \\ +\ 7 \\ \hline \end{array}$

4 Write the name and the number of edges.

a

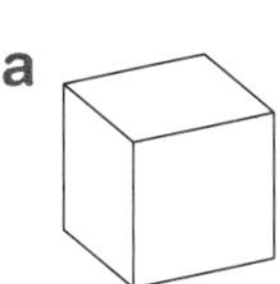

______

b

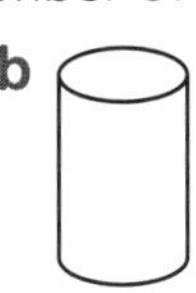

______

5

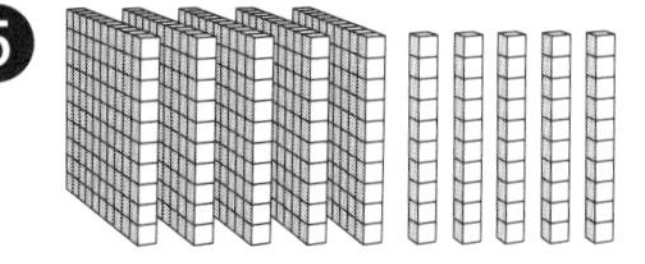

☐ hundreds	☐ tens	☐ ones

6 **a** 2, 4, 6, ______, ______, ______, ______

b 23, 33, 43, ______, ______, ______, ______

c 165, 155, 145, ______, ______, ______

7 Which of the terms *impossible*, *unlikely*, *likely* or *certain* describes the chance of you finding $20 tomorrow?

______

8 Write 20 in words. ______

9 Draw and label two 3D objects that can easily be stacked.

5:4 ☐ out of 7

Extension

1 **a** $46 + 25 =$ ______ **b** $58 + 34 =$ ______

2 $30 + 3 - 3 + 5 - 5 + 1 - 1$ ______

3 $100 - 2 - 2 - 2 - 2 - 2 - 2 - 2$ ______

4 $57 + 3 + 3 + 3 + 3 + 3 + 3 + 3$ ______

5 Complete this picture if the dotted line is a line of symmetry.

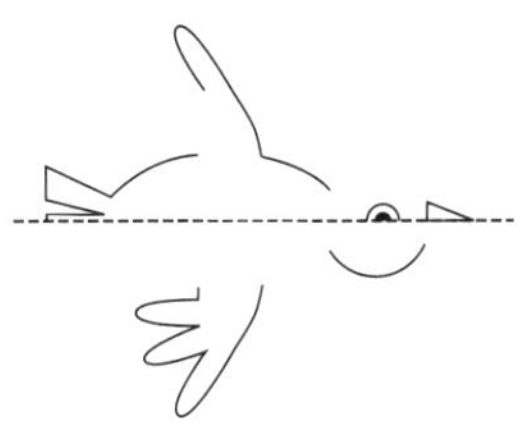

6 Write 342 in words. ______

______

7 **a** Are these numbers equally likely to be spun? ______

b How many 2s are most likely in 12 spins? ______

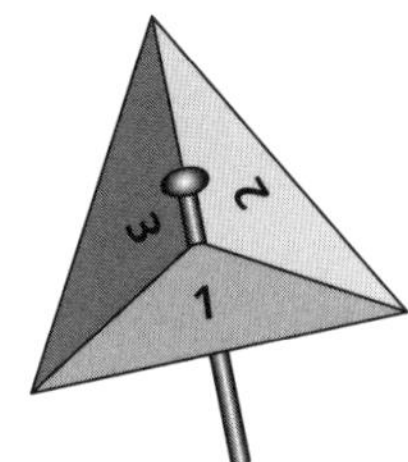

Challenge

Draw a picture that has symmetry.

Chance

Place the letter from these boxes on the line below, in order, from *least likely* to *most likely*.

The next person to walk through the classroom door will be:

A a teacher	**B** the principal	**C** my mother
D a human	**E** a father	**F** a boy

least likely ______________________ most likely

  ISBN 978 0 6557 0883 4

6:1   ☐ out of 18

1. 47 + 3 ____
2. 19 − 3 ____
3. 16 + ____ = 20
4. 27 + ____ = 30
5. $\begin{array}{r} 22 \\ +\ 17 \\ \hline \end{array}$
6. 3 × 10 ____
7. Digits in 451. ____
8. 12 ÷ 2 ____
9. 4 ÷ 2 ____
10. $\begin{array}{r} 38\text{c} \\ -\ 24\text{c} \\ \hline \end{array}$
11. 493 = ____ hundreds, ____ tens, ____ ones
12. Put these numbers in order from smallest to largest: 212, 129, 76, 318, 236.

    ____

13. Draw a line of symmetry on this picture.

14. Show 725 on this numeral expander.

15. The odd numbers between 10 and 20.

    ____, ____, ____, ____, ____

16. Which is large: fifty or fifteen? ____
17. Write 903 in words.

    ____

    ____

18. December is the ____ month of the year.

6:2   ☐ out of 16

1. 28 + 8 ____
2. 29 − 7 ____
3. 16 + 5 ____
4. 19 + 4 ____
5. $\begin{array}{r} 26 \\ +\ 12 \\ \hline \end{array}$
6. 4 × 10 ____
7. Digits in 324. ____
8. Digits in 32. ____
9. Double 23. ____
10. $\begin{array}{r} 42\text{c} \\ -\ 21\text{c} \\ \hline \end{array}$
11. 942 = ____ hundreds, ____ tens, ____ ones
12. How many fingers on 3 boys? ____
13. Order from smallest to largest.

    857, 758, 875   ____, ____, ____

14. The number modelled. ____

15. Finish these pictures using symmetry.

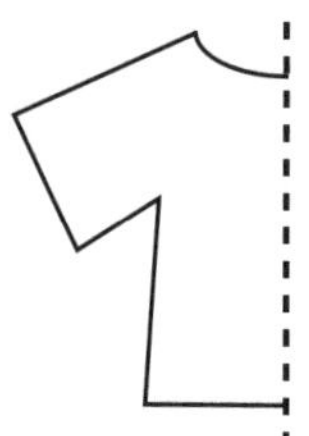

16.
    a Is 713 larger than 709? ____
    b Is 682 larger than 628? ____
    c Is 707 larger than 770? ____

Turn to ID card B on page 7.
Give the answers for these numbers.

(5) ____   (6) ____
(7) ____   (11) ____
(12) ____   (13) ____
(14) ____   (15) ____
(18) s ____   c ____

 ISBN 978 0 6557 0883 4

6:3 [ ] out of 9

1. 25 + 23
2. 47c + 24c
3. 73 + 15
4. 452 = _____ hundreds, _____ tens, _____ ones
5. Which shapes have only flat surfaces?

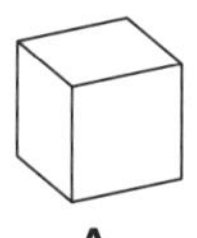

A

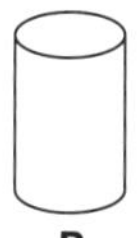

B

C

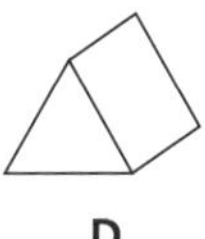

D

____

6. Use the jump strategy to find:

   a 36 + 7 = _______

   b 58 + 16 = _______

7. Draw the line of symmetry on this picture.

8.

9	4
13	

a 9 + 4 = ____ b 13 − 9 = ____

c 4 + 9 = ____ d 13 − 4 = ____

9. The total value of these notes. _______

6:4 Extension [ ] out of 7

1. Draw two lines of symmetry on this shape.

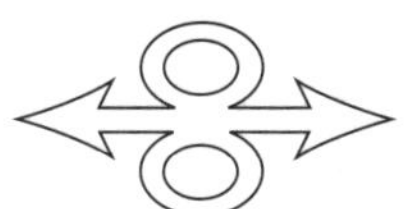

2. 9 + 1 + 8 + 2 + 10 + 0 _______
3. 40 − 4 − 4 − 4 − 4 − 4 _______
4. These ladybirds have the same number of dots on both sides. How many dots altogether? _______

5. If this wavy line were made 3 times as long, how many hills would be drawn? _______

6. Rachel saw the same number of koalas in 5 trees. There were 15 koalas. How many in each tree? _______
7. Show this number as tens and ones.

Challenge

Draw as many symmetrical letters, shapes and symbols as you can.

6 + 4 = [10], 4 + 6 = [10], 10 − 6 = [4], 10 − 4 = [6]

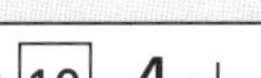
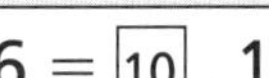
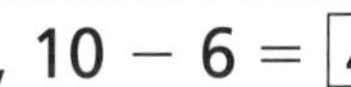

Complete four number sentences for each picture.

a 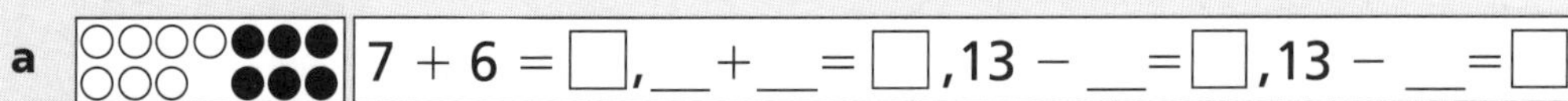

7 + 6 = [ ], __ + __ = [ ], 13 − __ = [ ], 13 − __ = [ ]

b 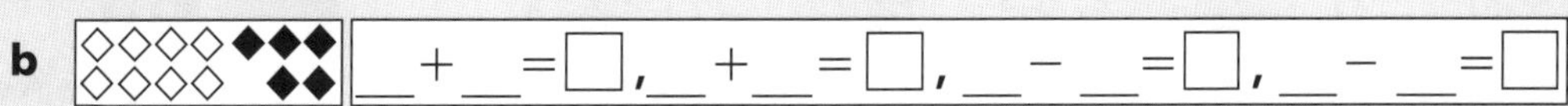

__ + __ = [ ], __ + __ = [ ], __ − __ = [ ], __ − __ = [ ]

c 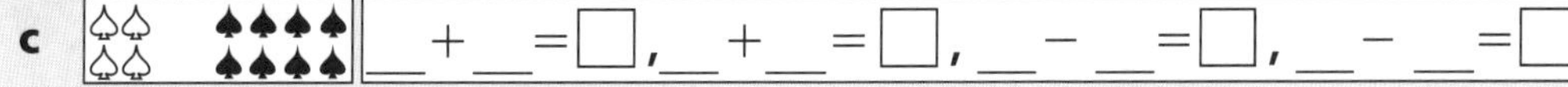

__ + __ = [ ], __ + __ = [ ], __ − __ = [ ], __ − __ = [ ]

7:1 — out of 16

1. 59 + 4 ____
2. 27 − 2 ____
3. 2 × 2 ____
4. 1 × 2 ____
5. $\begin{array}{r} 34 \\ +12 \\ \hline \end{array}$
6. 4 × 5 ____
7. 2 × 5 ____
8. 1 × 10 ____
9. 3 × 10 ____
10. $\begin{array}{r} 34\,c \\ -22\,c \\ \hline \end{array}$
11. 635 = ____ hundreds, ____ tens, ____ ones
12. Draw a line of symmetry.

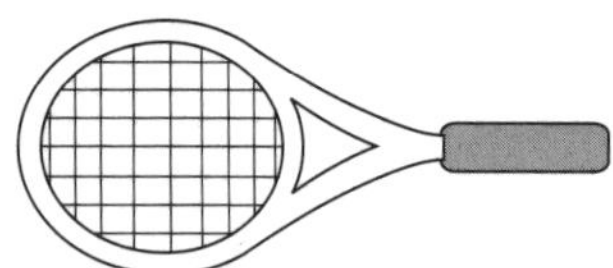

13. Write the value of each note.

____ ____ ____

14. A 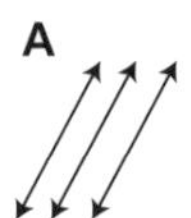 B 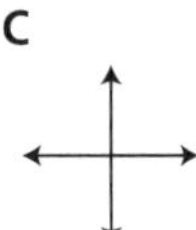 C

Which group shows parallel lines? ____

15. Join the dots.

This is a regular ____.

16. What shape is this cross-section of a cone? ____

7:2 — out of 19

1. 37 + 6 ____
2. 54 − 5 ____
3. 3 × 2 ____
4. 6 × 2 ____
5. $\begin{array}{r} 37 \\ +12 \\ \hline \end{array}$
6. 3 × 5 ____
7. 1 × 5 ____
8. 2 × 10 ____
9. 4 × 10 ____
10. $\begin{array}{r} 69\,c \\ -27\,c \\ \hline \end{array}$

11. 586 = ____ hundreds, ____ tens, ____ ones
12. How many corners has:

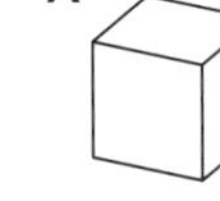

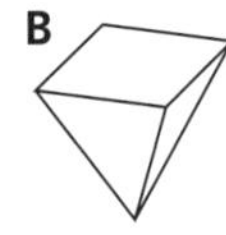

a A? ____

b B? ____

13. How many fingers on 8 girls? ____
14. The total value of these notes. ____

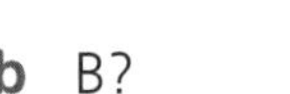

15. Draw a line of symmetry on each picture.

16.

7	8
15	

a 7 + 8 = ____ b 15 − 8 = ____

c 8 + 7 = ____ d 15 − 7 = ____

17. I have 2 bunches of 10 flowers.

How many flowers altogether? ____

18. Which is largest: 462, 264 or 459? ____
19. A triangle has ____ sides.

× tables

×2: 0, 5, 8, 2, 7, 4, 10, 3, 9, 6

×0: 0, 5, 8, 2, 7, 4, 10, 3, 9, 6

Rows of nothing.

7:3 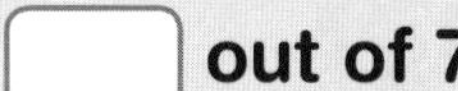 out of 7

1
$$\begin{array}{r} 37 \\ +\ 12 \\ \hline \end{array}$$

2
$$\begin{array}{r} 69c \\ -\ 27c \\ \hline \end{array}$$

3 **a** Name this shape. ____________

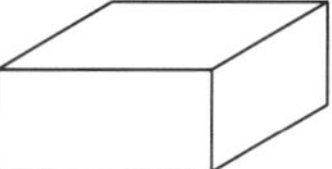

How many: **b** faces? ________

c corners? ________ **d** edges? ________

e What would the cross-section of this object be? ____________

4 5 + 9 = 14 and 9 + 5 = ________

Write the linking subtraction sentences.

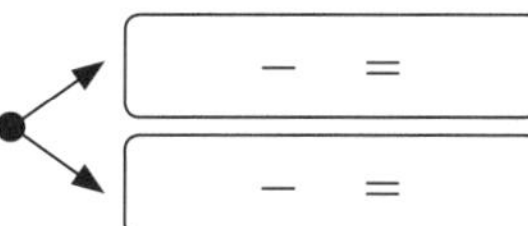

5 **a** How many fingers on 4 hands? ________

b 4 × 5 = ________

c How many toes on 5 children? ________

d 5 × 10 = ________

6 Which picture shows lines that are:

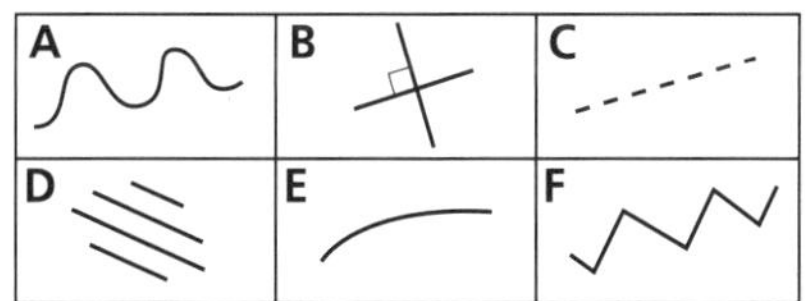

a parallel? _____ **b** perpendicular? _____

7 Which is largest: 550, 555 or 505? ________

7:4 Extension out of 4

1 Complete this picture if the broken line is a line of symmetry.

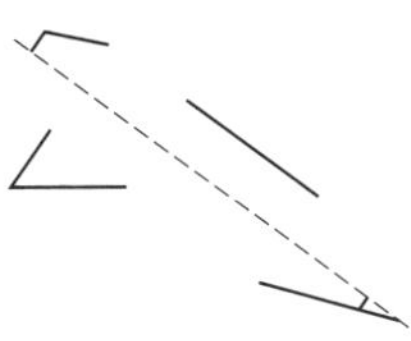

2 | 5 | hundreds | 3 | tens | 1 | ones |

= ________ tens and ________ one

3 If 2 non-parallel lines are drawn to cross these parallel lines, at how many points would lines cross altogether? ________

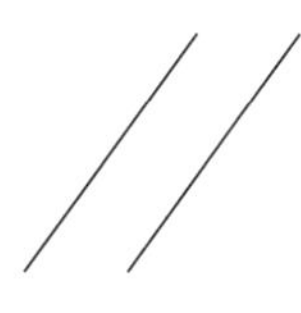

4 I played golf. What was my total score for these 4 holes? ________

Hole 1	8
Hole 2	9
Hole 3	7
Hole 4	10

Challenge

Draw a 3D object and write everything you know about it.

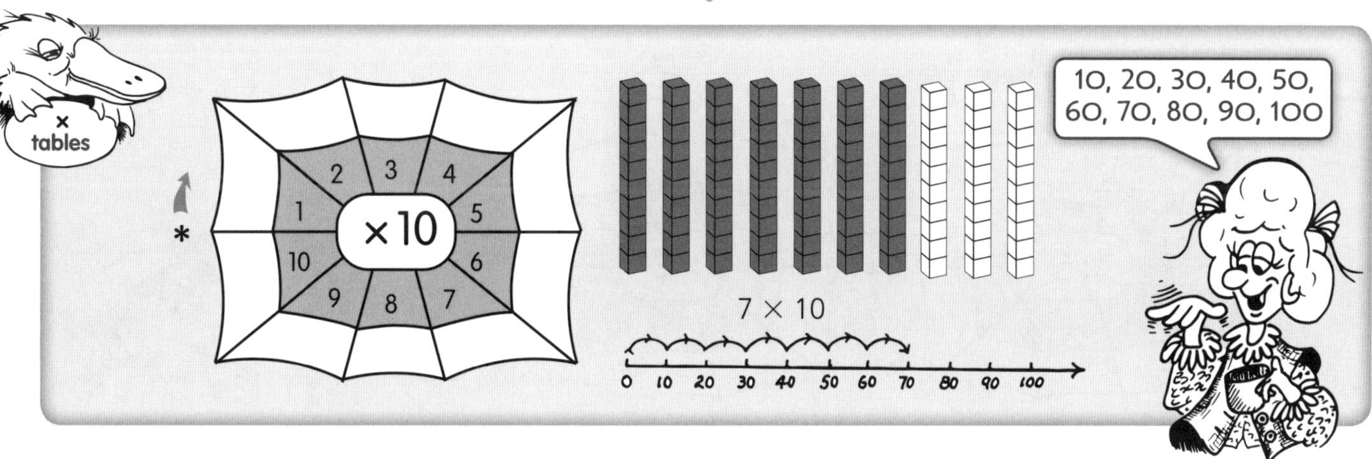

8:1 out of 17

1. 26 + 4 ______
2. 56 − 6 ______
3. 8 × 2 ______
4. 10 × 2 ______
5. $\begin{array}{r} 40 \\ +\ 11 \\ \hline \end{array}$
6. 10 × 5 ______
7. 2 × 5 ______
8. 6 × 10 ______
9. 7 × 10 ______
10. $\begin{array}{r} 42c \\ -\ 11c \\ \hline \end{array}$

11. 3 groups of 6. ______

12. These shapes are r______ shapes.

13.

	3	7	4	5	2	10	6	9
× 5								

14. Which is closer to 1 metre: the height of a cat, the width of a door or the length of a car? ______

15. 

______ rows of ______ = ______

16. Two weeks is also called a fortnight.
Days in 1 fortnight. ______

17. Begin at 90 and join the dots to count backwards by tens.

90 40 60 70 80 50

8:2 out of 16

1. 47 + 23 ______
2. 67 − 17 ______
3. 4 × 2 ______
4. 8 × 2 ______
5. $\begin{array}{r} 66 \\ +\ 31 \\ \hline \end{array}$
6. 8 × 5 ______
7. 6 × 5 ______
8. 5 × 10 ______
9. 10 × 10 ______
10. $\begin{array}{r} 87c \\ -\ 16c \\ \hline \end{array}$

11. Join the dots.
This is an irregular ______.

12. I swam 20 metres, then 30 metres and then 40 metres. How far did I swim altogether? ______

13. I have 63 birds.
How many more do I need so I have 77?
______ + ______ = ______

14. Draw two parallel lines.

15. Write 128 in words. ______

16. **a** Write the smallest 3-digit number possible using 0, 9 and 7. ______
b Write the largest 3-digit number possible using 9, 9 and 8. ______

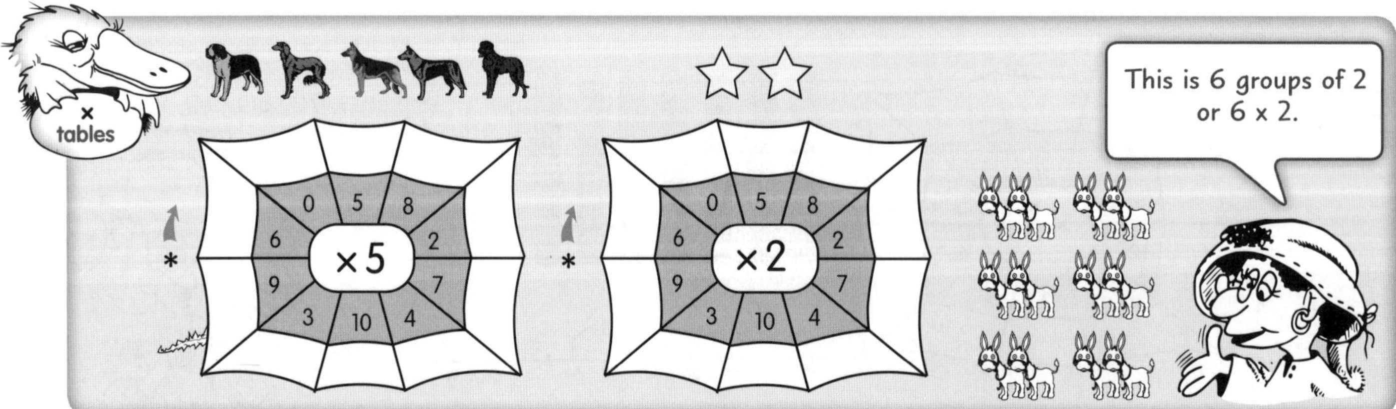

 AUSTRALIAN SIGNPOST MATHS 3 MENTALS • ISBN 978 0 6557 0883 4

8:3 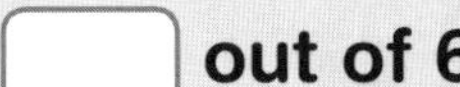 out of 6

1

```
  44
+ 23
```

2

```
  58c
- 34c
```

3 Use the jump strategy to find:

a 51 − 23 = _____

b 39 + 24 = _____

4
a 14 − 5 = _____ **b** 24 − 5 = _____
c 34 − 5 = _____ **d** 44 − 5 = _____
e 36 + 7 = _____ **f** 46 + 7 = _____
g 56 + 7 = _____ **h** 66 + 7 = _____

5 Which box shows:

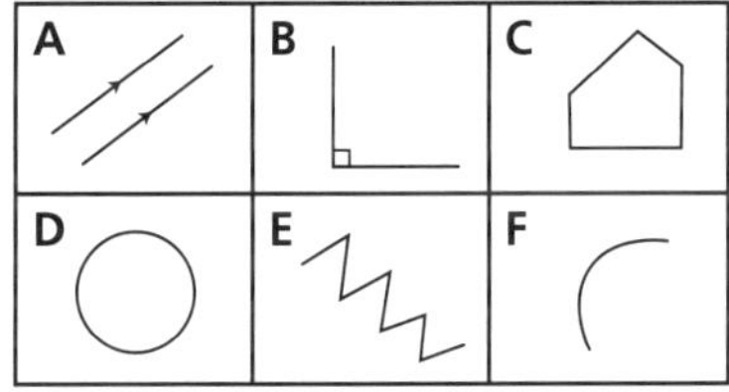

a a curved line? _______
b an irregular shape? _______
c parallel lines? _______
d a right angle? _______

6 Write the largest number possible using 1, 7 and 3. _______

8:4 Extension out of 5

1 How many place-value tens blocks, placed end-to-end, would reach:

a 1 metre? _______ **b** 2 metres? _______

2 **a** 17 + 9 = _______ **b** 35 + 9 = _______
c 32 − 9 = _______ **d** 45 − 9 = _______

3 I came 1st in a race out of 8. How many came:

a in front of me? _______
b behind me? _______

4 I will give Li two of these toys. How many different groups of 2 could I give? _______

5 Estimate the number of stars, circle groups of 10, then count. _______

Challenge

42 + 35 = _______

Explain how you found your answer.

____

____

____

Two 5s can be traded for one 10.

rows of 5

groups of 5

	2	4	6	8	10
× 5					

	1	2	3	4	5
× 10					

4 fives would make 2 tens.

 • *AUSTRALIAN SIGNPOST MATHS 3 MENTALS* • ISBN 978 0 6557 0883 4

9:1 out of 18

1. 32 + 6 ____
2. 42 + 6 ____
3. 52 + 6 ____
4. 62 + 6 ____
5. 40 + 10
6. 5 × 2 ____
7. 6 × 2 ____
8. 8 × 10 ____
9. 4 × 10 ____
10. 60c − 10c
11. 11 − 6 = ____ so 6 + ____ = 11

    The difference between 6 and 11 is ____.
12. Fourteen fish are to be put into two equal groups. How many fish in each group?

    14 ÷ 2 = ____
13. How many pens would fit along the length of your ruler? ____
14. 12:15 can also be called ____ past ____.
15. Minutes in 1 hour. ____
16. What time is shown?

    ____
17. 5 + ____ = 11
18. Show quarter past 4 on the clock face.

    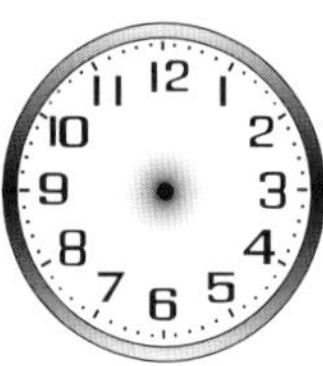

9:2 out of 16

1. 56 + 7 ____
2. 66 + 7 ____
3. 76 + 7 ____
4. 86 + 7 ____
5. 68 + 11
6. 7 × 2 ____
7. 7 × 5 ____
8. 6 × 5 ____
9. 9 × 10 ____
10. 69c − 10c
11. 18 − 13 = ____ so 13 + ____ = 18

    The difference between 13 and 18 is ____.
12. Is the length of a car closer to 2 metres, 4 metres or 20 metres? ____
13. Write the digital time.

    a  a quarter past 8

    b  half past 9
14. **a** Minutes in half an hour. ____

    b Minutes in one hour. ____
15. Each person at the table used 2 plates. How many plates were used if there were:

    a 6 people at the table? ____

    b 9 people at the table? ____
16. Use the jump strategy to find:

    a 38 + 57 = ____

    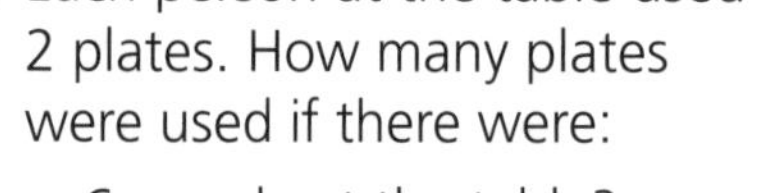

    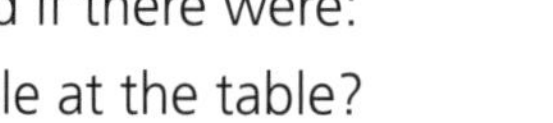

    b 42 − 27 = ____

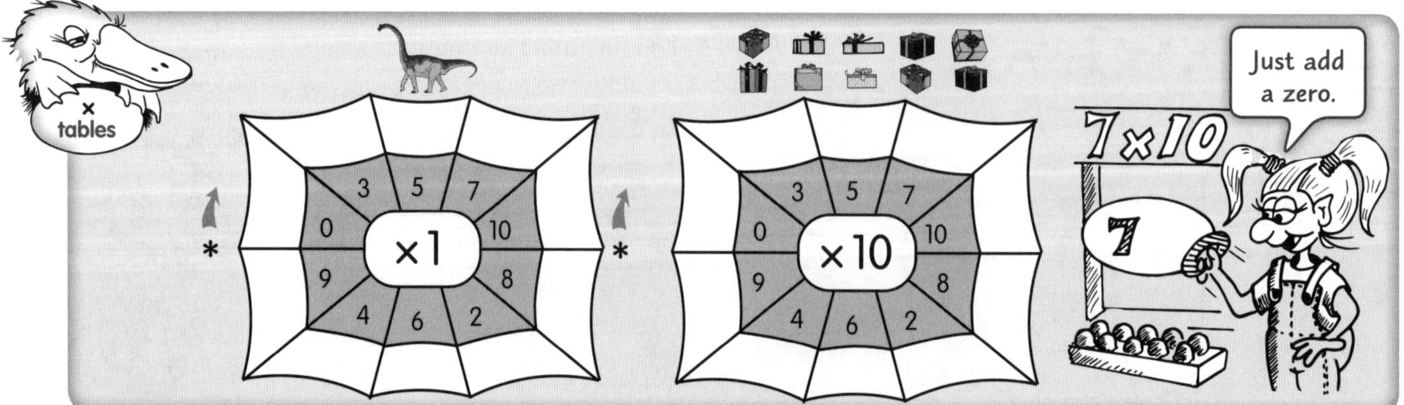

 ISBN 978 0 6557 0883 4

9:3 ☐ out of 10

1 $\begin{array}{r} 75 \\ +\ 13 \\ \hline \end{array}$

2 $\begin{array}{r} 54c \\ -\ 12c \\ \hline \end{array}$

3 Aziz saved his pocket money for five weeks. If he got \$2 a week, how much did he save? ______

4 12:45 means ______ past ______ or ______________ to ______.
Show 8:45 on this clock face.

5 22 − 17 = ______ so 17 + ______ = 22

The difference between 17 and 22 is ______.

6 Use the jump strategy to find:

a 46 + 39 = _____

b 51 − 36 = _____

7 5 fives take away 2 fives = ______________

8 How many wheels on 3 cars?

3 groups of ______ = ______ 3 × 4 = ______

9 What is half of 16? ______

10 Days in 2 weeks? _____

9:4 ☐ out of 7

Extension

1 76 − 27 = ______ so 27 + ______ = 76
The difference between 27 and 76 is ______.

2

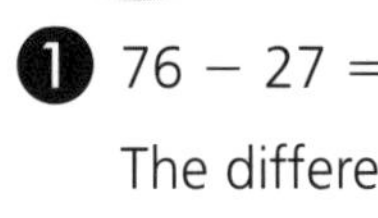

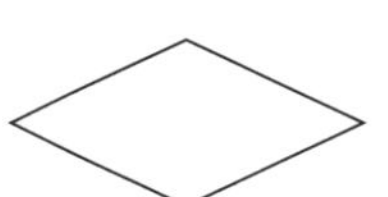

How many sides are on 3 rectangles and 4 rhombuses? ______

3 Is the length of a 30 cm ruler closer to one quarter or one half of a metre? ______

4 Which is larger: 3 × 5 or 8 + 8? ______

5 How many lines of symmetry has a regular hexagon? ______

6 **a** 3 rows of 5 plus 4 rows of 3 ______

b 2 rows of 5 plus 6 rows of 5 ______

c 2 rows of 7 plus 2 rows of 9 ______

7 How many hexagons would be in row:

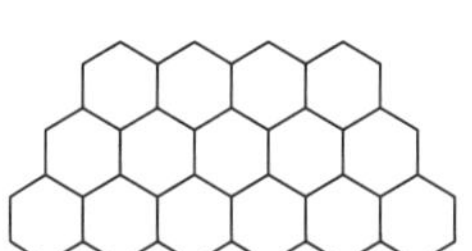

a 5? ______

b 10? ______

Challenge

Write as many number sentences as you can linking 40, 16 and 24.

____

____

____

Concept

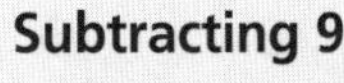

Subtracting 9

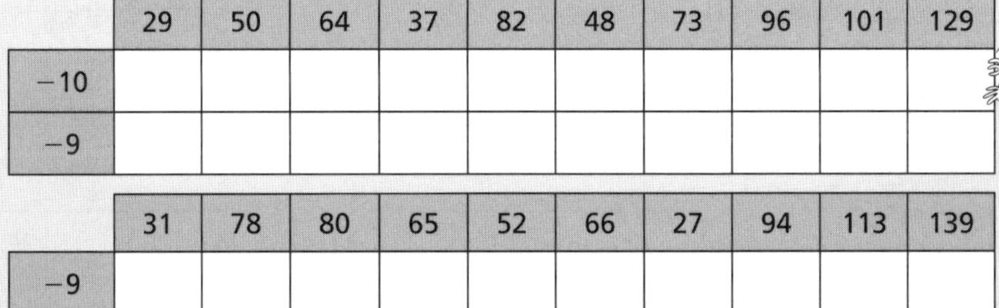

	29	50	64	37	82	48	73	96	101	129
−10										
−9										

	31	78	80	65	52	66	27	94	113	139
−9										

To subtract 9, take away 10 and add 1.

 • *AUSTRALIAN SIGNPOST MATHS 3 MENTALS* • ISBN 978 0 6557 0883 4

10:1 ☐ out of 17

1. 29 + 6 ____
2. 67 − 7 ____
3. 3 × 2 ____
4. 2 × 2 ____
5.
```
  30
+ 12
```
6. 3 × 5 ____
7. 1 × 5 ____
8. 2 × 10 ____
9. 4 × 10 ____
10.
```
  53c
− 11c
```

11. 14 − 8 = ______ so 8 + ______ = 14

    The difference between 8 and 14 is ______.
12. Circle the abbreviation for metres.

    A M **B** ms **C** m **D** MS
13. Write the digital time.

    a 6 o'clock [:] **b** 10 o'clock [:]

14. I have 3 trays of cookies. There are 10 cookies on each tray.

    How many cookies altogether? ______
15. 50c + 20c + 10c ______
16. Which container holds about 1 litre?

    A   **B** 

17. Use the jump strategy to find:

    26 + 8 = ____

10:2 ☐ out of 19

1. 38 + 13 ____
2. 62 − 13 ____
3. 4 × 2 ____
4. 8 × 2 ____
5.
```
  43
+ 26
```
6. 4 × 5 ____
7. 8 × 5 ____
8. 9 × 10 ____
9. 10 × 10 ____
10.
```
  67c
− 34c
```

11. Are you taller than 1 metre? ______.
12. 29 − 15 = ______ so 15 + ______ = 29

    The difference between 15 and 29 is ______.
13. Fingers on two hands. ________
14. **a** 12 − 6 = ____ **b** 22 − 6 = ____

    c 32 − 6 = ____ **d** 42 − 6 = ____

    e 34 + 9 = ____ **f** 44 + 9 = ____

    g 54 + 9 = ____ **h** 64 + 9 = ____
15. How many 5c coins make 50c? ______
16. I have 34 cards.

    How many more do I need so I have 42?

    ______ + ______ = ______
17. 2:30 means ______ past ____.

    Show 2:30 on this clock face.
18. The abbreviation for 1 metre? ________
19. What holds more:

    a mug or a 1 litre bottle? __________

Addition linked with subtraction

Concept

a

7 − 4 = ___	4 + ___ = 7

The difference is ____.

b

10 − 8 = ___	8 + ___ = 10

The difference is ____.

c

13 − 5 = ___	___ + 5 = 13

The difference is ____.

d

28 − 12 = ___	12 + ___ = 28

The difference is ____.

e

34 − 12 = ___	12 + ___ = 34

The difference is ____.

f

44 − 36 = ___	36 + ___ = 44

The difference is ____.

10:3 out of 10

1. 55 + 33

2. 50c − 30c

3. 67 + 31

4. I have 6 rows of 10 trees.

   How many trees altogether? _______

5. Is the height of a door greater than 1m? _______

6. Write the digital time.

   a  half past 7

   b  quarter to 6

7. 8:45 means _______ past _______ or _______ to _______.

   Show 8:45 on this clock face.

8. 32 − 23 = _______ so 23 + _______ = 32

   The difference between 23 and 32 is _______.

9. Use the jump strategy to find:

   a 47 + 35 = _____

   b 53 − 25 = _____

10. How many wheels on 5 cars? _______

    5 groups of _______ = _______

    5 × 4 = _______

10:4 Extension out of 5

1. Chloe spent $2 each day for two fortnights.

   How much did she spend? _______

2. 5 sheep and 6 chickens.

   How many legs? _______

3. Colour the coins that are equal in value to twenty-two 10 cent coins.

4. What is the total value of three $2.00 coins and three 50c coins? _______

5. Find the difference between the total of the 2 smallest-value coins and the total of the 2 largest-value coins. _______

Challenge

Explain different ways you can make $3, e.g. $1 + $1 + 50c + 20c + 20c + 10c.

____

____

____

____

____

____

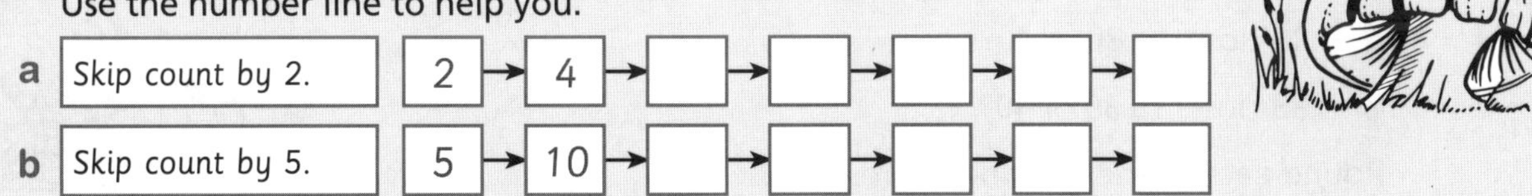

In each case, skip count, writing in the numbers as you go.
Use the number line to help you.

a Skip count by 2. 2 → 4 → ☐ → ☐ → ☐ → ☐ → ☐

b Skip count by 5. 5 → 10 → ☐ → ☐ → ☐ → ☐ → ☐

c Skip count by 10. 10 → ☐ → ☐ → ☐ → ☐ → ☐ → ☐ → ☐ → ☐

11:1 ☐ out of 18

1. 13 + 5 ____
2. 23 + 5 ____
3. 33 + 5 ____
4. 43 + 5 ____
5. 50 + 30 = ____
6. 10 × 2 ____
7. 5 × 2 ____
8. 2 × 10 ____
9. 5 × 10 ____
10. 70c − 20c = ____
11. 12 metres minus 6 metres. ____
12. Which coin is worth more than 20c but less than $1? ____
13. I bought a cake for 10c and an iceblock for 50c. How much did I spend? ____
14.

Container	Estimate	Measure
Jug	8 cups	7 cups
Tub	28 cups	33 cups

How many cups does:

a the jug hold? ____

b the tub hold? ____

15. Circle the numbers that round off to 60.

54 61 67 69 51 56
55 59 64 58 65

16. 563 = ____ hundreds, ____ tens, ____ ones
17. Double 6 then halve your answer. ____
18. **a** Share 10 between 2. ____ each

b Share 10 between 5. ____ each

11:2 ☐ out of 18

1. 48 + 6 ____
2. 38 + 6 ____
3. 28 + 6 ____
4. 18 + 6 ____
5. 87 + 12 = ____
6. 6 × 2 ____
7. 8 × 2 ____
8. 8 × 5 ____
9. 9 × 5 ____
10. 93c − 61c = ____
11. Is the width of an elephant about 2 metres? ____
12. **a** How many 5 cent coins? ____

b What is the total value? ____

13. 20c + 5c + 10c + 50c ____
14. Circle the numbers that round off to 400.

321 412 392 351
456 333 406 349

15. The total value of these notes. ____

16. 482 = ____ hundreds, ____ tens, ____ ones
17. 345, 355, 365, ____, ____, ____
18. **a** 265 + 17 − 17 = ____

b 534 + 21 – 21 = ____

Rounding a number to the nearest ten

If it ends in 5, 6, 7, 8 or 9, round up.

If it ends in 1, 2, 3 or 4, round down.

a Is 36 closer to 30 or 40? ____

b is 33 closer to 30 or 40? ____

Round each of these to the nearest ten.

c 44 ____ **d** 17 ____ **e** 85 ____

 • *AUSTRALIAN SIGNPOST MATHS 3 MENTALS* • ISBN 978 0 6557 0883 4

11:3 ☐ out of 9

1. $\begin{array}{r} 64 \\ +34 \\ \hline \end{array}$

2. $\begin{array}{r} 94c \\ -32c \\ \hline \end{array}$

3. $\begin{array}{r} 40 \\ +52 \\ \hline \end{array}$

4. Is your teacher taller than 1 metre? ________

5. Circle the numbers that round off to 800.

851	809	751	793	783
824	750	850	849	

6. 904 = _____ hundreds, _____ tens, _____ ones

7. **a** Colour half. **b** Colour two quarters.

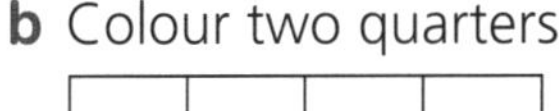

c Colour four eighths.

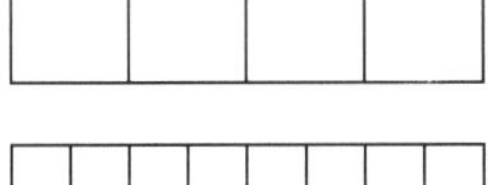

8. **a** Colour $\frac{3}{8}$ of the circles.

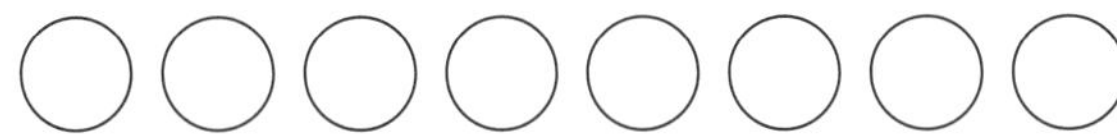

b Colour $\frac{5}{8}$ of the rectangles.

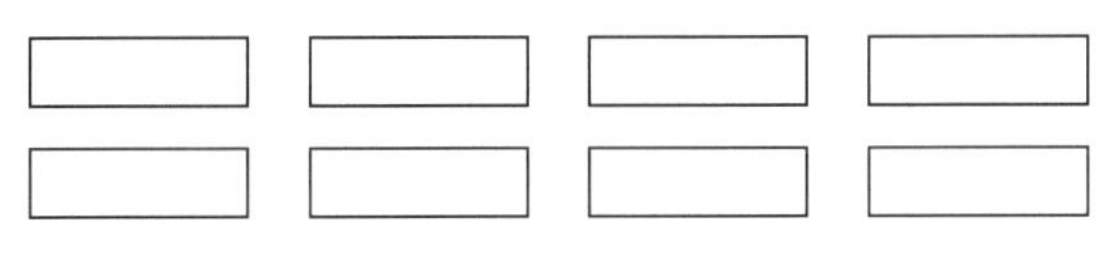

c What fraction of the rectangles are not coloured in? ☐

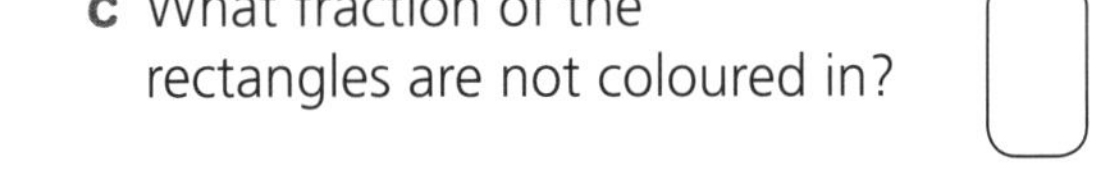

9. **a** 2, 4, 6, _____, _____, _____, _____, _____

b 10, 20, 30, _____, _____, _____, _____, _____

c 23, 33, 43, _____, _____, _____, _____, _____

11:4 ☐ out of 6

Extension

1. How many 5c coins make $2? __________

2. How many 10c coins could I get from:

a $2.10? __________ **b** $4.20? __________

3. If this pattern were repeated, what would the 8th shape be? __________

4. How many different ways can you order these pictures in a row? __________

5. Halve 84 then halve your answer. __________

6. 100 years = 1 c __________

Challenge

Draw and label containers that hold more than 1 litre.

When you play noughts and crosses, in how many different ways can you make:

a a line of three crosses? ________

b two noughts, side by side? ________

12:1   ☐ out of 16

1. 6 + ____ = 13
2. 8 + ____ = 14
3. 9 + ____ = 15
4. Digits in 354. ____
5. $\begin{array}{r} 40 \\ +\ 40 \\ \hline \end{array}$
6. 5 × 5 ____
7. 5 × 2 ____
8. 5 × 10 ____
9. 3 × 5 ____
10. $\begin{array}{r} 90c \\ -\ 10c \\ \hline \end{array}$
11. 935 = ____ hundreds, ____ tens, ____ ones
12. Circle the numbers that round off to 70.

74	71	78	63	69	73
68	65	66	75	67	

13. cm stands for ____________________.

    m stands for ____________________.

14. A B C

    a Which is the longest path? ________

    b Which is the shortest path? ________

15. Colour:

    a $\frac{1}{3}$ 

    b $\frac{1}{4}$

    c $\frac{1}{5}$ 

    d $\frac{1}{10}$

16. 473 = ____ hundred, ____ tens, ____ ones

12:2   ☐ out of 16

1. 16 + ____ = 25
2. 19 + ____ = 27
3. 35 + 15 ____
4. 25 + 65 ____
5. $\begin{array}{r} 46 \\ +\ 32 \\ \hline \end{array}$
6. 8 × 2 ____
7. 8 × 5 ____
8. 8 × 10 ____
9. 6 × 5 ____
10. $\begin{array}{r} \$56 \\ -\ \$50 \\ \hline \end{array}$
11. Centimetres in 1 metre. ________
12. True or false? Your fingernail is about 1 cm long. ________
13. Which line is:

    a 2 cm long? ____

    b 3 cm long? ____

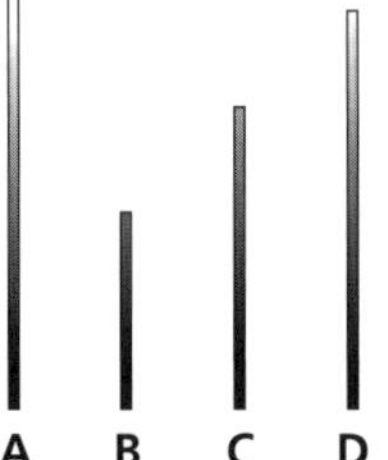

14. The coloured part is ____ out of ____. $\frac{\square}{9}$

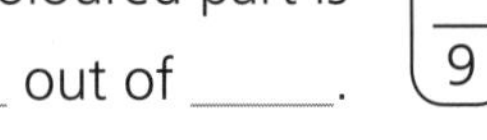

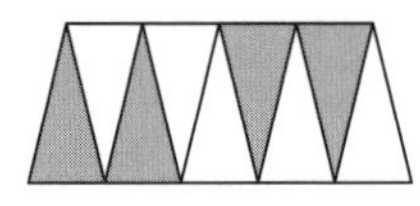

15. Circle the numbers that round off to 300.

328	349	250	241	249
351	240	261	350	

16. Circle 7 tenths. $\frac{\square}{\square}$

Rounding to the nearest hundred

Look at the tens digit to see if you should round up or down.

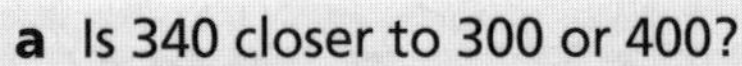

a Is 340 closer to 300 or 400? ________

b Is 653 closer to 600 or 700? ________

Round each of these to the nearest 100.

c 339 ________ d 814 ________ e 751 ________ f 490 ________ g 119 ________

Round up if the tens digit is 5, 6, 7, 8 or 9.

Round down if the digit is 1, 2, 3 or 4.

 • *AUSTRALIAN SIGNPOST MATHS 3 MENTALS* • ISBN 978 0 6557 0883 4

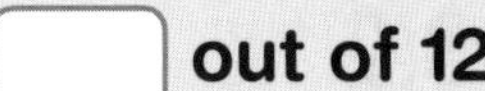

12:3 ☐ out of 12

1. 34 + 35
2. $46 − $31
3. 53 + 25
4. What part is shaded? ______ out of ______ $\frac{\square}{\square}$

5. Is a giraffe taller than 1 metre? __________
6. Centimetres in 1 metre. __________
7. Measure the width of this page to the nearest centimetre. __________cm
8. Circle the numbers that round off to 900.

937	949	950	957	849
851	840	961	850	

9. Which shapes will roll? ____________________

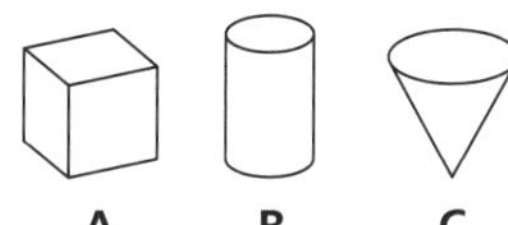

10. **a** 2, 4, 6, ______, ______, ______, ______
    b 3, 6, 9, ______, ______, ______, ______
    c 5, 10, 15, ______, ______, ______, ______
    d 10, 20, 30, ______, ______, ______
11. This shows ______ groups of ______.

 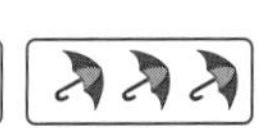 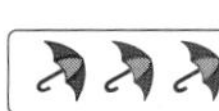

12. 2 × 3 = ______ 3 × 3 = ______ 4 × 3 = ______

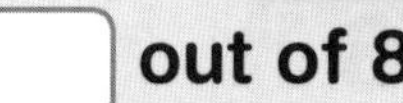

12:4 ☐ out of 8 **Extension**

1. 1 metre − 1 centimetre = __________
2. 50 cm is **half / a quarter / an eighth** of a metre.
3. Quarters in 6 apples. __________
4. **a** Half of 12 = ______ **b** A quarter of 12 = ______
5. How many 2 cm lengths could be cut from the ribbon below? __________
6. 

   What fraction is shaded?

   A $\frac{4}{20}$ **B** $\frac{5}{10}$ **C** $\frac{1}{4}$ ______
7. How many different ways can you order these pictures in a row? __________

8. **a** 5 pentagons have ______ sides.
   b 5 hexagons would have ______ sides.

Challenge

Write numbers that would round to 700.

____

____

____

____

Writing fractions

The part shaded is:

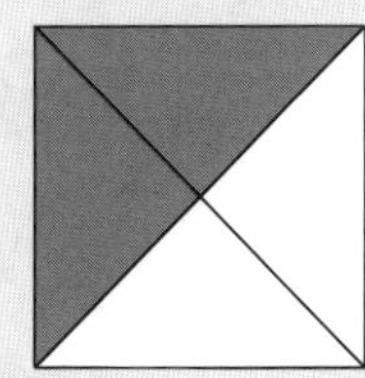

2 out of 4 or $\frac{2}{4}$

a 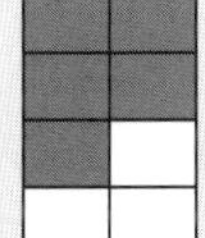 ______ out of ______ or $\frac{\square}{\square}$

b 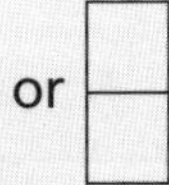 ______ out of ______ or $\frac{\square}{\square}$

c  ______ out of ______ or $\frac{\square}{\square}$

d ______ out of ______ or $\frac{\square}{\square}$

13:1 out of 16

1. 40 + 40 ____
2. 50 − 20 ____
3. 3 × 2 ____
4. 4 × 2 ____
5. $\begin{array}{r} 30 \\ +\ 40 \\ \hline \end{array}$
6. 3 × 5 ____
7. 4 × 5 ____
8. 3 × 10 ____
9. 4 × 10 ____
10. $\begin{array}{r} 50c \\ -\ 40c \\ \hline \end{array}$
11. Each half is: [|]
    ____ out of ____ equal parts.
12. 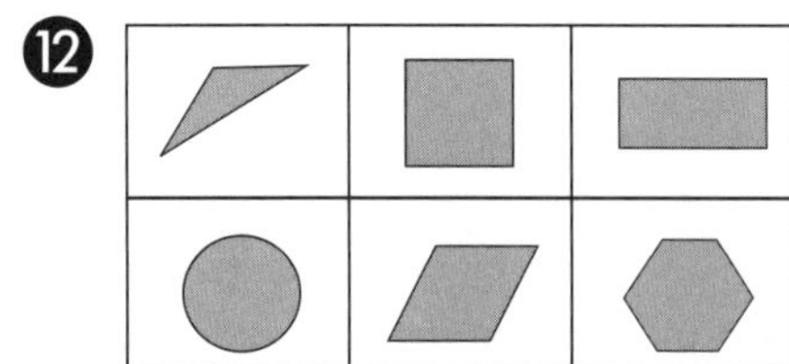

    What shape is at the:
    a bottom left? ____
    b top right? ____
13. Would you use centimetres (cm) or metres (m) to measure the length of a rabbit? ____.
14. Is the width of your finger about 1 cm? ____
15. What part is shaded? ____ out of ____

16. **a** 6 + ____ = 10 **b** 5 + ____ = 10
    c 7 + ____ = 10 **d** 2 + ____ = 10
    e 9 + ____ = 10 **f** 4 + ____ = 10

13:2 out of 16

1. 34 + 25 ____
2. 39 − 34 ____
3. 5 × 2 ____
4. 6 × 2 ____
5. $\begin{array}{r} 55 \\ +\ 34 \\ \hline \end{array}$
6. 5 × 5 ____
7. 6 × 5 ____
8. 5 × 10 ____
9. 6 × 10 ____
10. $\begin{array}{r} 70c \\ -\ 30c \\ \hline \end{array}$
11. Write 1 metre 65 centimetres using the short form. ____
12. Find the distance around the shape. ____
    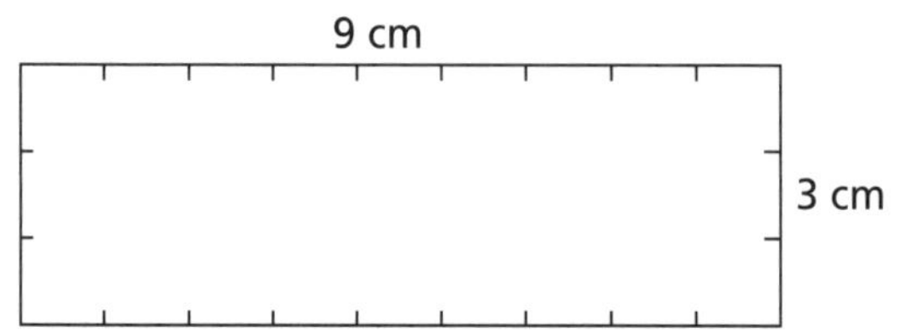

13. What part is shaded?
    ____ out of ____

14. A stove is less than 15 cm high. True or false? ____.

15. Measure each line in centimetres.
    A ____
    B ____
    C ____
16. Is 32 closer to 30 or 40? ____

Using number lines

−4 −10 −2

19 20 21 22 23 24 25 26 27 28 29 30 31 32 33 34 35 36 37 38 39 40 41 42

tables

Jump to the nearest ten to add or subtract. These jumps show 42 – 16 = 26.

a 66 – 27 = ____ **b** 52 – 15 = ____ **c** 71 – 36 = ____

66 52 71

13:3 ☐ out of 12

1

tens	ones
2	2
+1	1

2

tens	ones
2	4
+2	3

3 Round 54 to the nearest 10. ______

4 Measure the height of this page to the nearest centimetre. ______

5 Write 4 metres and 27 centimetres using the short form. ______

6 Which is bigger: one quarter or one half? ______

7 Is the length of this about 1 cm, 3 cm or 5 cm? ______

8 Would you use centimetres (cm) or metres (m) to measure the height of a cup? ______

9 6 days and 8 days.

How many days altogether?

____ + ____ = ____

10 Is the width of your finger about 1 cm? ______

11 Use the jump strategy to find:

47 + 35 = ______

12 Which is larger: 492, 941 or 914? ______

13:4 ☐ out of 6

Extension

1 Centimetres in 3 metres. ______

2 **a** Draw a line to show the halfway point.

b Draw 2 more lines to make quarters.

c Colour one quarter of this group.

3 How many halves in 2 oranges? ______

4 **a** Colour half of this shape.

b How many quarters in one half? ______

c What is larger: $\frac{1}{2}$ or $\frac{1}{4}$? ______

5 **a** What part is shaded?

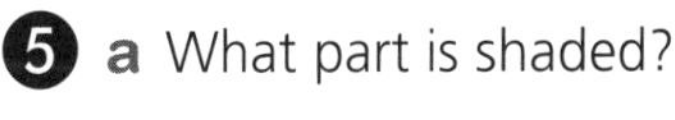

____ out of ____

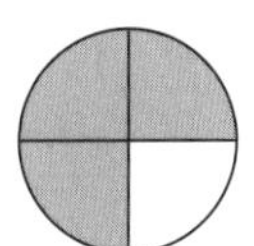

b 3 quarters plus 1 quarter. ______

6 Use this picture to fill in the boxes.

a ☐ − 6 = ☐

b ☐ − 9 = ☐

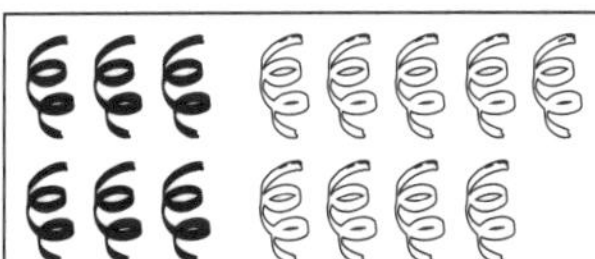

Challenge

Use the jump strategy to find 43 − 28.

+	5	10	4	0	2	6	1	3
2								
3								
4								
5								
10								

 • *AUSTRALIAN SIGNPOST MATHS 3 MENTALS* • ISBN 978 0 6557 0883 4

14:1 out of 16

1. 10 × 2 ____
2. 15 − 3 ____
3. 11 + 6 ____
4. 11 − 3 ____
5. 30 + 40
6. Add 4 and 7. ____
7. 6 minus 2. ____
8. Total 5 and 3. ____
9. 9 shared by 3. ____
10. 50c − 40c

11. Would you use centimetre (cm) or metres (m) to measure the length of a pool? ____
12. Circle the duck 2nd from the right.

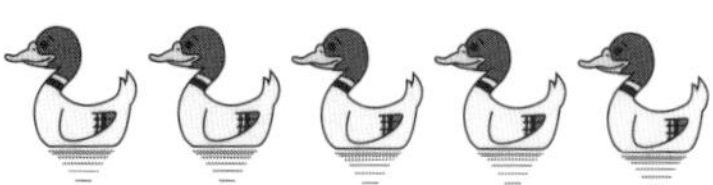

13. Use the jump strategy to find:

59 + 18 = ____

14. Use mental strategies to find:

a 300 + 200 = ____ b 18 + 6 + 2 = ____

c 32 + 21 = ____ d 19 + 9 = ____

15. I bought 32 apples and 16 were eaten.

____ − ____ = ____

How many are left? ____

16. a 10, 12, 14, ____, ____, ____, ____

b 10, 15, 20, ____, ____, ____, ____

c 36, 46, 56, ____, ____, ____, ____

14:2 out of 17

1. 23 + 61 ____
2. 56 − 51 ____
3. 41 + 42 ____
4. 45 − 21 ____
5. 62 + 13
6. 3 × 5 ____
7. 7 × 2 ____
8. 9 × 10 ____
9. 4 × 2 ____
10. 95c − 62c

11. Going up from **A**, where do I finish if I turn 2nd on the left, then 1st right, then right then keep going?

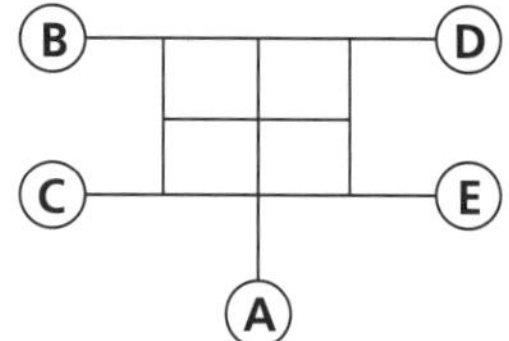

____

12. Write 6 metre 93 centimetres using the short form. ____
13. Write 1 m 35 cm in centimetres. ____ cm
14. Follow the directions and colour the path of the counter.

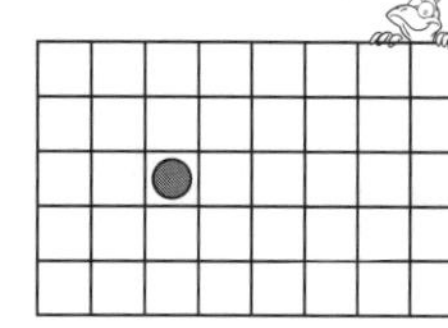

Move 2 down, then 4 right, then 3 up.

15. Use the jump strategy to find:

34 − 16 = ____

16. Use the compensation strategy to find:

a 38 + 19 = ____ b 27 + 19 = ____

17. We had 34 apples and ate 10. How many apples do we have now? ____

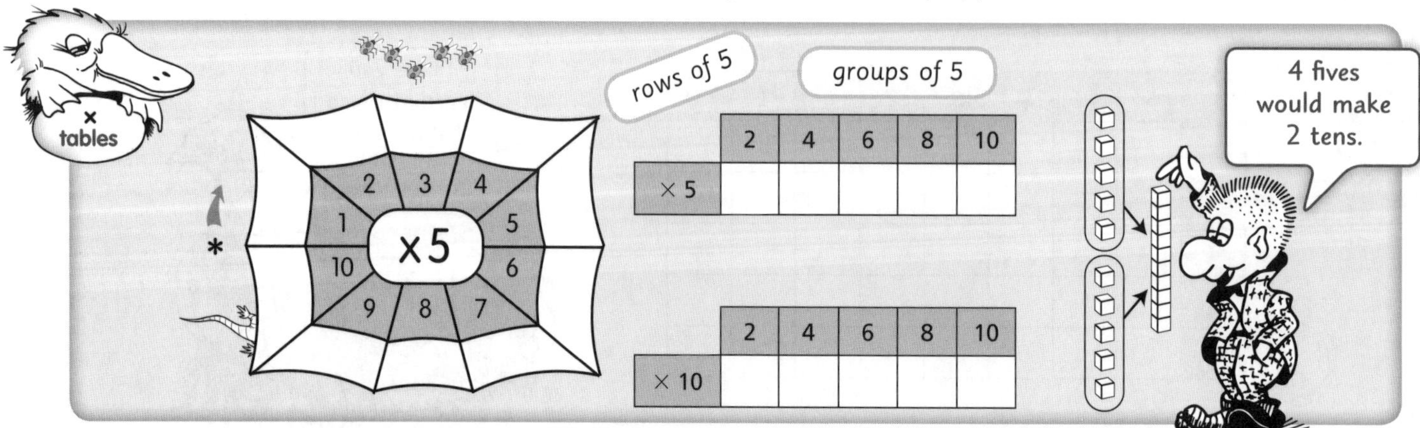

	2	4	6	8	10
× 5					

	2	4	6	8	10
× 10					

 ISBN 978 0 6557 0883 4

14:3 out of 5

1

tens	ones
8	1
+1	3

2

tens	ones
$4	3
+$3	4

3 Use the compensation strategy to find:

a 45 + 19 = ______ **b** 34 + 29 = ______

c 68 + 26 = ______ **d** 83 − 19 = ______

e 76 − 39 = ______ **f** 54 − 18 = ______

4

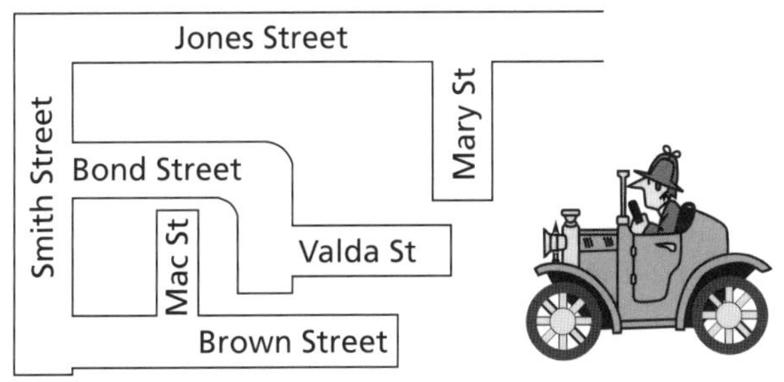

a Mac Street joins ______

b Mary Street joins ______

c Valda Street joins ______

d What Street joins Jones Street and Brown Street? ______

e What other streets do you travel on to get from Valda Street to Mac Street?

______

______

5 Use the jump strategy to find:

61 − 24 = ______

14:4 Extension out of 6

1 Centimetres in 7 metres. ______

2 2 metres minus 50 cm. ______

3 Quarters in 5 pears. ______

4 Where does the ant stop if he goes:
2 up, 3 left, 1 down,
5 right and then 3 up? ______

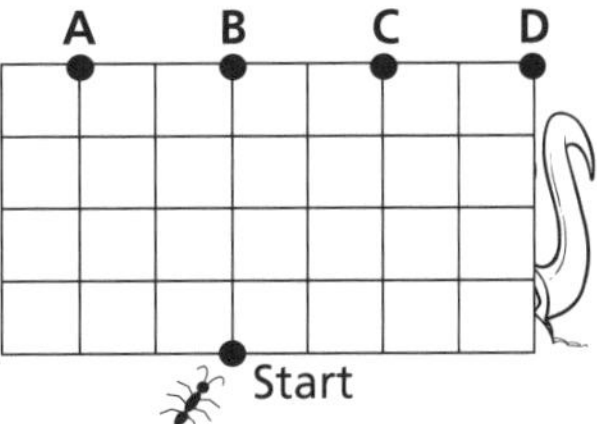

5 True or false?

a 4 × 5 = 2 × 10 ______

b 8 × 5 = 4 × 10 ______

6 Circle the correct number sentence for this picture.

A 2 × 5 = 10 **C** 12 − 2 = 10

B 2 × 7 = 14 **D** 7 + 5 = 12

Challenge

39 + 25 = ☐

Explain how you found your answer.

______

______

______

Find the distance around each shape.

a 5 cm, 2 cm
Distance = ______ cm

b 4 cm, 2 cm
Distance = ______ cm

c 10 cm, 1 cm
Distance = ______ cm

Add all the sides to get the distance.

15:1 ☐ out of 15

1. 23 + 10 ______
2. 22 − 10 ______
3. 20 + 30 ______
4. 32 − 14 ______
5. $\begin{array}{r} 24 \\ +\ 24 \\ \hline \end{array}$
6. 8 × 2 ______
7. 5 × 5 ______
8. 7 × 10 ______
9. 9 × 10 ______
10. $\begin{array}{r} 35c \\ -\ 13c \\ \hline \end{array}$
11. The number modelled is __________.

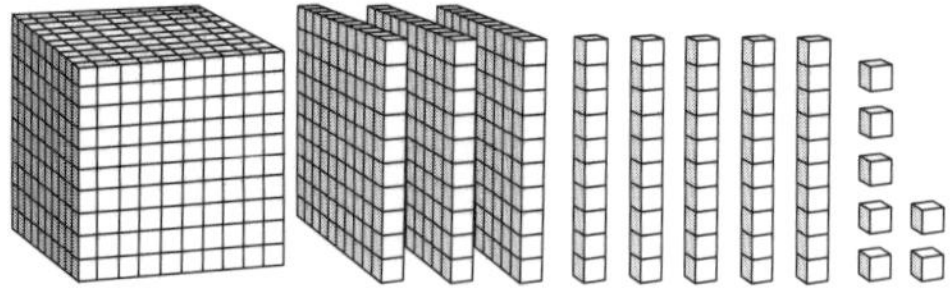

12. Is the width of a door about 1 metre? __________.
13. Can we measure height in metres, litres or grams? __________.
14. **Favourite Colours**

Red	Blue	Orange	Yellow
10	12	4	7

   a The most popular colour was __________.
   b The least popular colour was __________.
   c How many people liked blue? __________
   d How many people liked red? __________
15. a 300 + 400 = ________
   b 500 + 200 = ________
   c 400 + 400 = ________

15:2 ☐ out of 15

1. 45 + 10 ______
2. 62 − 10 ______
3. 300 + 500 ______
4. 38 + 19 ______
5. $\begin{array}{r} 35 \\ +\ 34 \\ \hline \end{array}$
6. 9 × 2 ______
7. 9 × 5 ______
8. 8 × 10 ______
9. 10 × 10 ______
10. $\begin{array}{r} 46c \\ -\ 15c \\ \hline \end{array}$
11. Is the length of your foot longer than 40 cm? __________
12. Write the numeral shown on the abacus. __________

13. Write 4 m 72 cm in centimetres. __________ cm
14. **Swimming Attendance**

	Day 1	Day 2	Day 3	Day 4	Day 5	Day 6	Day 7	Day 8
Yuri		●		●	●	●		●
Sue	●	●	●		●			
Tan	●	●	●		●	●		●
Jess	●	●	●	●	●	●	●	
Greg	●	●	●	●	●	●	●	●
Mari	●		●	●	●	●	●	●

   a Who was never away? __________
   b Who was away the most? __________
   c How many days was Jess present? __________
15. How many more than 10 is 19? __________

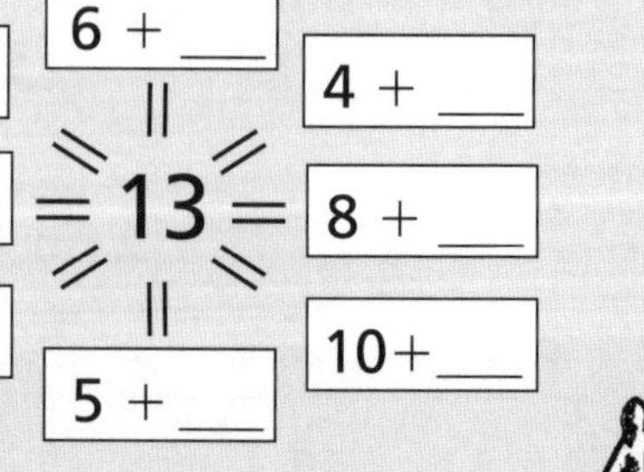

Finish the boxes.

 • *AUSTRALIAN SIGNPOST MATHS 3 MENTALS* • ISBN 978 0 6557 0883 4

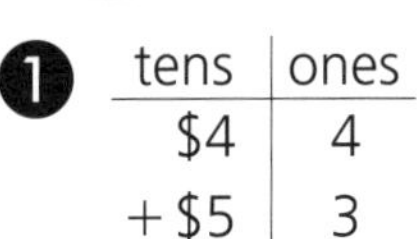

15:3 out of 10

1

tens	ones
$4	4
+ $5	3

2

tens	ones
$6	3
+ $2	1

3 The number modelled is ______________.

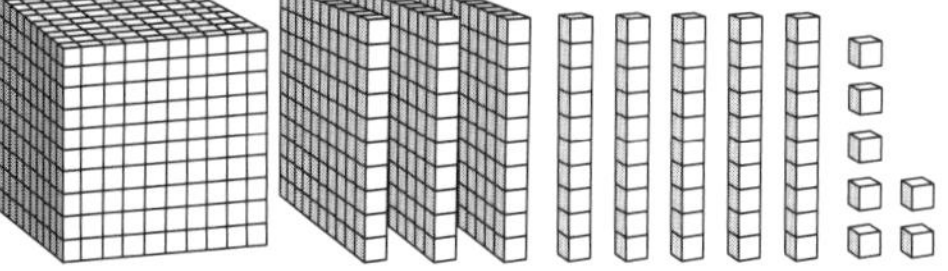

4 How many digits in 4302? __________

5 Write the numeral for five thousand, four hundred and twenty-three. __________

6 Bridge to the next 10 to find:

a 47 + 6 = _______ b 28 + 7 = _______

7 Use the jump strategy to find:

a 46 + 27 = _______

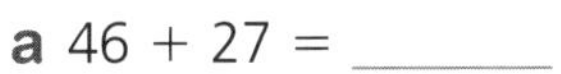

46

b 73 − 27 = _______

73

8 Use the compensation strategy to find:

a 46 + 19 = ______ b 38 + 29 = ______

9 Is 5192 odd or even? __________

10 a 5000 + 300 + 20 + 9 = __________

b 9000 + 500 + 10 + 7 = __________

15:4 out of 8

Extension

1 Quarters in 10 oranges. __________

2 Write 3 even numbers that are larger than 4203 and less than 4214.

____

3 How many faces are on 10 cubes? __________

4 In a race I am 7th out of 10. How many:

a are in front of me? __________

b are behind me? __________

5 How many cakes will be in the 16th row of this pattern?

____

6 Write three different 1-digit numbers that add to give 24. __________

7 Does 4 × 5 = 2 × 10? __________

8 Write the next odd number after 6029. __________

Challenge

84 − 29 = ☐

Explain how you found your answer.

____

____

____

____

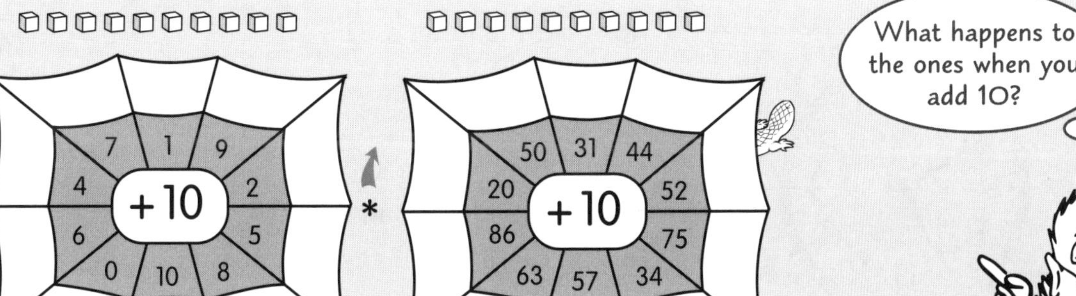

16:1 out of 16

1. 300 + 100 ______
2. 600 − 100 ______
3. 35 + 21 ______
4. 45 + 10 ______
5. 44 + 44
6. Digits in 193. ______
7. Half of 16. ______
8. Double 7. ______
9. 6×5 ______
10. 67c − 20c

11. **Favourite Animal**

Bird	Dog	Cat	Mouse
7	13	4	1

a The most popular animal was a ____________.

b The least popular animal was a ____________.

c How many people liked dogs? ____________

d How many people liked birds? ____________

12. Write the numeral six thousand, five hundred and seventy-two. ____________

13. Show twenty-two past 10 on both clocks.

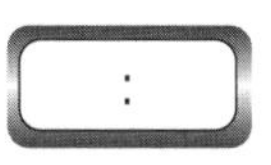

14. Circle the largest number.
5309, 5911, 5400

15. How many digits in 5299? ____________

16. a September is the _____ month of the year.

b July is the _____ month of the year.

16:2 out of 15

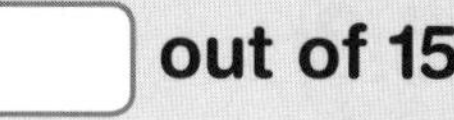

1. 400 + 500 ______
2. 800 − 500 ______
3. 67 + 24 ______
4. Half of 28. ______
5. 58 + 31
6. Digits in 100. ______
7. Half of 18. ______
8. Double 9. ______
9. 8 groups of 5. ______
10. 58c − 50c

11. What fraction is shaded? $\frac{\square}{\square}$

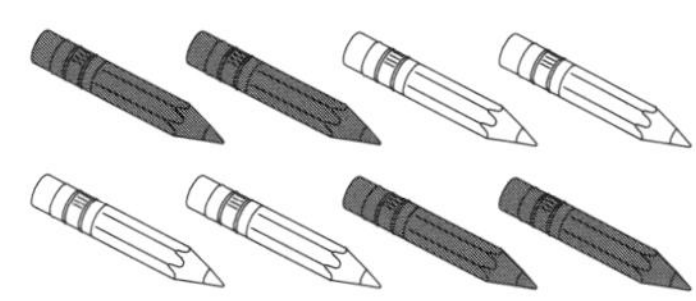

12. Write 4501 in words.

____

____

13. **Lollipops Eaten** (one symbol = one lollipop)

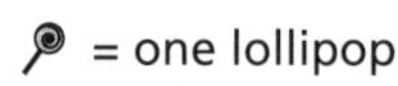

Leah	🍭🍭🍭🍭🍭🍭🍭🍭🍭🍭🍭
Robert	🍭🍭🍭🍭🍭🍭
Chen	🍭🍭🍭🍭🍭🍭🍭
Anna	🍭🍭

a Who ate the most lollipops? ____________

b Who ate the least lollipops? ____________

c How many did Chen eat? ____________

d How many lollipops were eaten altogether? ____________

14. 3523 = ____ thousands, ____ hundreds, ____ tens, ____ ones

15. Is 5290 odd or even? ____________

Measure

60 min / 0 min, 5 min, 10 min, 15 min, 20 min, 25 min, 30 min, 35 min, 40 min, 45 min, 50 min, 55 min

How many minutes are in:

a one hour? ____________

b half an hour? ____________

c a quarter of an hour? ____________

d three quarters of an hour? ____________

For this clock, how many minutes is it:

e after 6 o'clock? ____________

f before 7 o'clock? ____________

16:3 ☐ out of 9

1. 200 + 500

2. 600 + 300

3. 100 + 800

4. Show the time 11 minutes to 1.

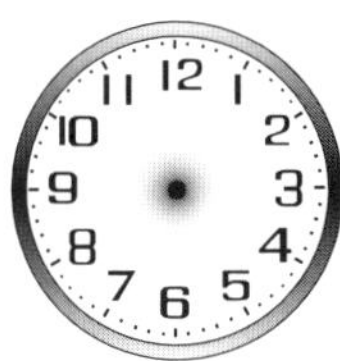

5. What number is shown here? ____________

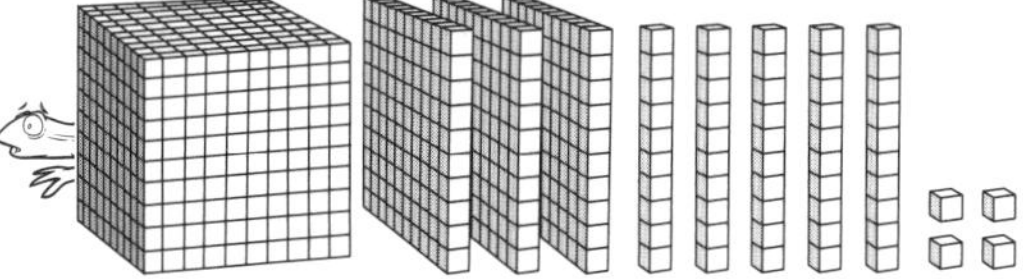

6. Colour $\frac{3}{8}$ of this shape.

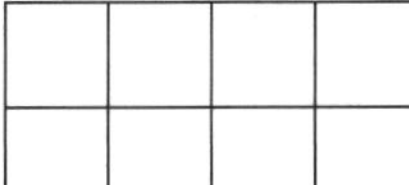

7. Circle the largest number.
   6999, 7020, 7011, 7001

8. Use this picture to complete the graph.

Kangaroos	Emus	Rabbits

9. I collected 31 cards and Jo collected 23.
   a How many did we collect altogether? _______
   b How many more did I collect than Jo? _______

16:4 ☐ out of 8

Extension

1. a 13 minutes after 10:38 is ____________.
   b 18 minutes before 12:16 is ____________.

2. How many blocks would be in the 6th row? _________

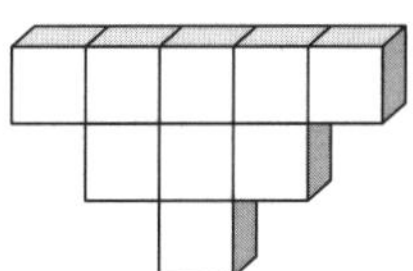

3. Circle $\frac{1}{3}$.

4. Wings on 14 owls. ____________

5. 60 − 10 − 5 − 5 − 10 ____________

6. How many eggs in 3 dozen? ____________

7. 73 was doubled and that answer was halved. What was the result? ____________

8. What fraction of an hour is:
   a 15 minutes? ____________
   b 30 minutes? ____________
   c 45 minutes? ____________

Challenge

- *Colour:*

one third *one eighth*

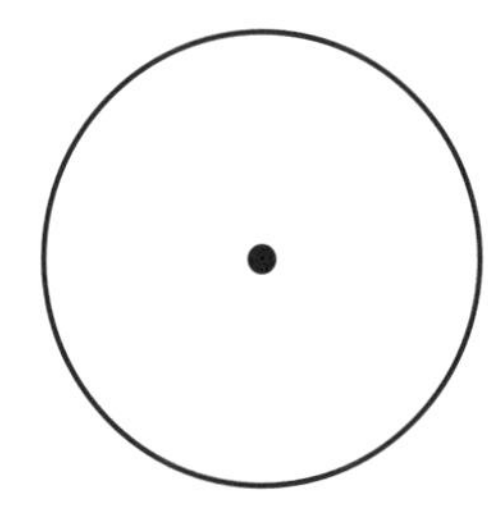

- *Which fraction is larger?* ____________

Concept

Our numerals Roman numerals

Use these clocks to give the Roman numeral for:

a 4 _______ **b** 5 _______ **c** 9 _______

d 10 _______ **e** 11 _______ **f** 12 _______

Write our numeral for:

g IX _______ **h** XI _______ **i** VI _______

17:1 out of 17

1. 3 × 2 ____
2. 6 ÷ 2 ____
3. 2 × 5 ____
4. 10 ÷ 2 ____
5. $\begin{array}{r} 44 \\ +\ 44 \\ \hline \end{array}$
6. Digits in 193. ____
7. Half of 16. ____
8. Double 7. ____
9. 6 × 5 ____
10. $\begin{array}{r} 67c \\ -\ 20c \\ \hline \end{array}$
11. What is the abbreviation for kilogram? ________
12. Would your shoes weigh more or less than a kilogram? ________
13. How many sides on:
    a 2 triangles? ________
    b 3 triangles? ________

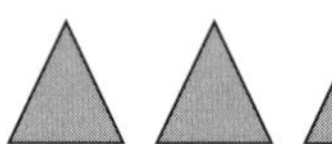
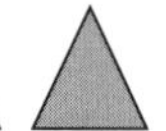

14. Colour one sixth of the fish.

15. What is the digital time? ________

16. Circle the correct abbreviation.
    4 KG,   4 kg,   4 Kg,   4 Kgs,   4 kgs
17. 345, 445, 545, ________, ________, ________

17:2 out of 17

1. 3 × 5 ____
2. 5 × 5 ____
3. 4 × 5 ____
4. 38 + 12 ____
5. $\begin{array}{r} 56 \\ +\ 32 \\ \hline \end{array}$
6. 15 ÷ 5 ____
7. 10 ÷ 5 ____
8. 12 ÷ 2 ____
9. 29 + 23 ____
10. $\begin{array}{r} 53c \\ -\ 40c \\ \hline \end{array}$
11. True or false? 1 litre of water has a mass of 1 kilogram. ________
12. What is the time shown on this clock face? ____ past ____

13. Use the short form to write 3 kilograms. ________
14. Write the numeral for the number shown on this abacus. ________

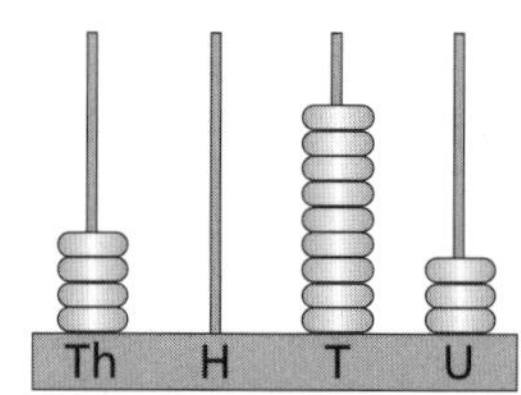

15. Write the time that is three minutes before:
    a 6 to 7 ________
    b 18 to 11 ________
16. What part is shaded? ____ out of ____ $\frac{\ }{\ }$
17. 3, 6, 9, ____, ____, ____, ____, ____

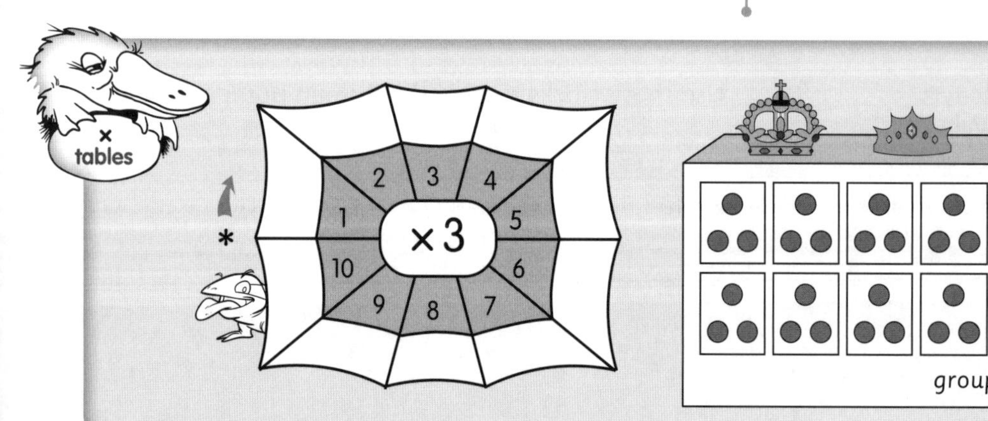

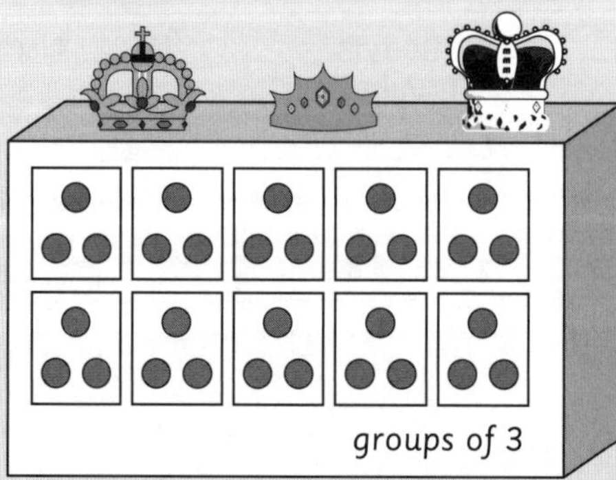

 • *AUSTRALIAN SIGNPOST MATHS 3 MENTALS* • ISBN 978 0 6557 0883 4

17:3 ☐ out of 9

1. Show each time on the clock.

   a 3:24

   b 12:47

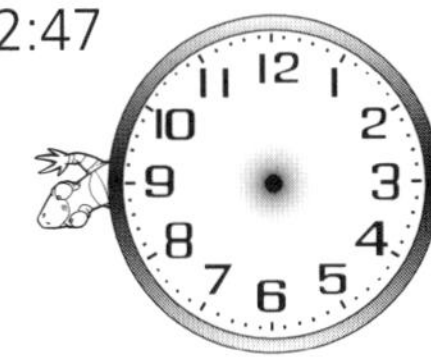

2. More or less than a kilogram:

   a this book? ________

   b your desk? ________

3. What analog time is shown?

   ____

4. Write the time that is four minutes after 4:53. ________

5. 

   The number modelled here is ________.

6. Would you use metres (m) or centimetres (cm) to measure the length of a truck? ________

7. Write 3 metres 35 centimetres in short form. ________

8. Write in words: 14, 40, 44.

   ____

   ____

9. **a** 87, 77, 67, ____, ____, ____, ____, ____

   b 30, 28, 26, ____, ____, ____, ____, ____

17:4 ☐ out of 8

Extension

1. How many different odd pairs could I make from 3 odd socks? ________

2. Circle $\frac{1}{4}$.

3. True or false?

   $\frac{1}{4}$ is more than $\frac{1}{8}$. ________

4. **a** Colour 5 sixths red and 1 sixth blue.

   b 5 sixths + 1 sixth ________

5. Twelve bolts have the same mass as 2 shoes. How many bolts have the same mass as 4 shoes? ________

6. Write the time that is 1 hour and 3 minutes before 4:36. ________

7. 12 △ 8 = 20 △ = ____

8. Would 6 pencils weigh more than 3 maths textbooks? ________

Challenge

Draw and label items that weigh less than 1 kg.

0 5 10 15 20 25 30 35 40 45 50 55

In each case, skip count, writing in the numbers as you go.
Use the number line to help you.

Skip count by 3. 3 → 6 → ☐ → ☐ → ☐ → ☐ → ☐

Skip count by 4. 4 → 8 → ☐ → ☐ → ☐ → ☐ → ☐

Skip count by 5. 5 → ☐ → ☐ → ☐ → ☐ → ☐ → ☐ → ☐ → ☐

18:1

out of 19

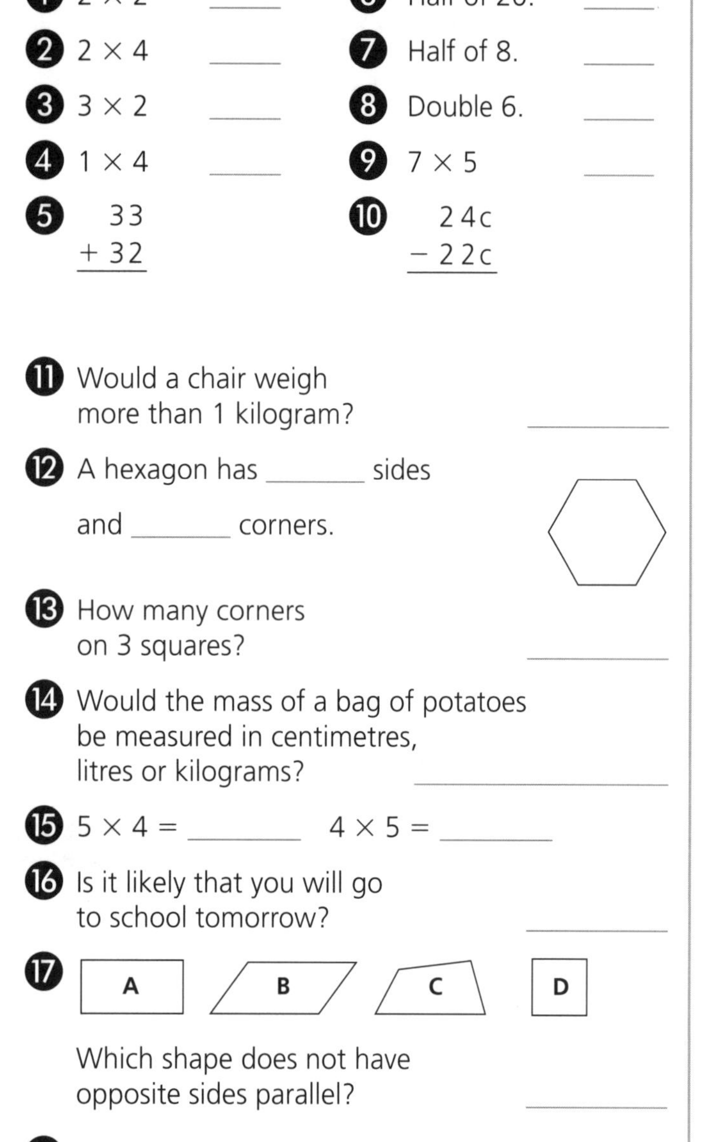

1. 2×2 ______
2. 2×4 ______
3. 3×2 ______
4. 1×4 ______
5. $\begin{array}{r} 33 \\ +\ 32 \\ \hline \end{array}$
6. Half of 20. ______
7. Half of 8. ______
8. Double 6. ______
9. 7×5 ______
10. $\begin{array}{r} 24c \\ -\ 22c \\ \hline \end{array}$
11. Would a chair weigh more than 1 kilogram? ______
12. A hexagon has ______ sides and ______ corners.
13. How many corners on 3 squares? ______
14. Would the mass of a bag of potatoes be measured in centimetres, litres or kilograms? ______
15. $5 \times 4 =$ ______ $4 \times 5 =$ ______
16. Is it likely that you will go to school tomorrow? ______
17. A B C D

    Which shape does not have opposite sides parallel? ______
18. I had 14 cards and lost some. I had 8 left. How many did I lose? ______
19. Draw 3 groups of 5 balls. $3 \times 5 =$ ______

18:2

out of 17

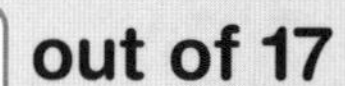

1. 8×10 ______
2. $80 \div 10$ ______
3. 8×2 ______
4. $16 \div 2$ ______
5. $\begin{array}{r} 55 \\ +\ 22 \\ \hline \end{array}$
6. 3×4 ______
7. 4×4 ______
8. 8×5 ______
9. 9×5 ______
10. $\begin{array}{r} 45c \\ -\ 24c \\ \hline \end{array}$
11. Use the short form to write 8 kilograms. ______
12. How many wheels on 4 cars? ______
13. Name the shape:

    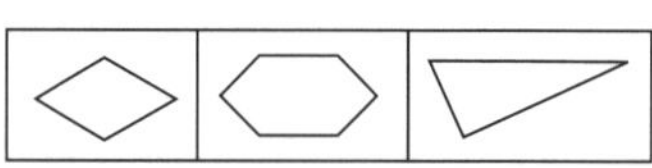

    a in the middle. ______

    b on the left. ______
14. How many animals in:

    a 32 pairs? ______

    b 43 pairs? ______

    c 50 pairs? ______

    d 21 pairs? ______
15. Which quadrilaterals have all sides equal?

    ______
16. Is it certain that any bird can fly? ______
17. a 75, 65, 55, ______, ______, ______, ______, ______

    b 3, 6, 9, ______, ______, ______, ______, ______

A coin is to be tossed 10 times.

a How many *heads* would be most likely? ______

b Toss a coin 10 times and show the results in this graph. Colour a circle for each toss.

Heads	○ ○ ○ ○ ○ ○ ○ ○
Tails	○ ○ ○ ○ ○ ○ ○ ○

Answers

ID card answers

ID card A

1 metres **2** centimetres **3** millimetres **4** litres **5** kilograms **6** minutes **7** centimetre **8** centimetre **9** even numbers **10** odd numbers **11** ordinal numbers **12** digits **13** times (or multiply) **14** 6 × 2 **15** division or divide **16** division or divide **17** equal **18** number line **19** number sentence **20** numeral expander **21** abacus **22** table **23** picture graph **24** tally **25** column graph **26** sector graph **27** digital watch **28** analog clock **29** calendar **30** balance scales

ID card B

1 angle **2** right angle **3** parallel lines **4** perpendicular lines **5** triangle **6** square **7** rectangle **8** diamond or rhombus **9** trapezium **10** parallelogram **11** quadrilaterals **12** pentagons **13** hexagons **14** circle **15** oval **16** regular shapes **17** irregular shapes **18** side, corner **19** line of symmetry **20** net of a cube **21** corner **22** edge **23** face **24** cube **25** prism **26** pyramid **27** base **28** cylinder **29** cone **30** sphere

1:1

1 **a** 64 **b** 82 **c** 36 **d** 46 **e** 108 **f** 120 **2** 10 **3** equal **4** 6, 8 **5** 7 **6** number line **7** 2 **8** 9 **9** 12

1:2

1 10 **2** 20 **3** 15 **4** 13 **5** 26 **6** 3 **7** $14 **8** 13, 15 **9** 10 **10** 15 **11** 8 – <u>4</u> = <u>4</u> **12** Half of the rectangle will be coloured.

13 **a** <u>half</u> past <u>8</u> (or <u>30</u> past <u>8</u>) **b** <u>quarter</u> past <u>1</u> (or <u>15</u> past <u>1</u>) **14** Four birds will be circled. **15** 4

Activity

3, 11, 4, 8, 12, 6, 10, 5, 9, 7

odd + odd = even

even + odd = odd

1:3

1 14 **2** 17 **3** 19 **4** 33 **5** 22 **6** 8, 10 **7** 26 **8** 19 **9** nineteen **10** **a** 25 **b** 23 **11** 28 **12** 52 **13** January **14** **a** 25 **b** 49 **c** 96 **d** 75

15 quarters, One quarter will be coloured. 

16 4th **17** 48, 50 **18** One pelican will be coloured.

19 55c

1:4

1 6 **2** 42, 32, 22 **3** 12, 15, 18, 21

4 (10 + 30) + (5 + <u>8</u>) = <u>40</u> + <u>13</u> = <u>53</u> **5** 3

6 20, 12, 8 **7** 4, 2 and 1

8 90 days (91 days in a leap year)

9 Summer (except in a leap year where Summer and Spring both have 91 days) **10** 8

Challenge

Answers will vary.

Activity

Answers will vary.

2:1

1 **a** 85 **b** 120 **c** 111 **d** 121 **e** 49 **f** 46 **2** **a** 20 **b** 20

3 30 **4** 18 **5** 7 **6** 3, Each has <u>3</u>. **7** 30

2:2

1 20 **2** 14 **3** 3 **4** 2 **5** 17 **6** 8 **7** 10 **8** 5 **9** 4 **10** 15c

11 One part of the circle will be coloured. 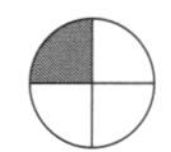

12 **a** 83 **b** 78 **c** 97 **d** 68

13 **a** 56 **b** 92

14 15, 15 **15** **A** rectangle **B** hexagon **C** trapezium

16 4 groups of 3 balls will be drawn. 12

Activity

13, 9, 6, 4, 10, 12, 8, 5, 11, 7

15, 11, 8, 6, 12, 14, 10, 7, 13, 9

odd + odd = <u>even</u>

2:3

1 46 **2** 40 **3** 26 **4** 47 **5** 26 **6** 11 **7** 66 **8** 40

9 **a** 53 **b** 73

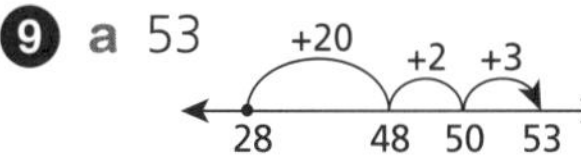

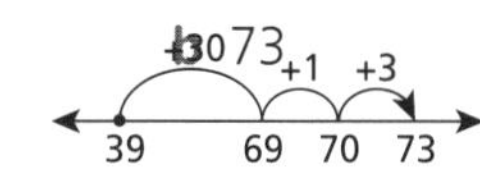

10 **a** 8 **b** 3 groups of <u>4</u> = <u>12</u> apples **11** **a** 3 **b** 3

12 13, <u>13</u> – <u>4</u> = <u>9</u> <u>13</u> – <u>9</u> = <u>4</u>

13 The fish will be circled.

2:4

1 **a** 8 **b** 8 **2** **a** 3 **b** 1 **3** BE KIND

4 quarter to 4 or 3:45

5 4 small rectangles will be coloured. **6** 30

Challenge

Answer will vary, e.g. 10 + 5, 3 × 5, 18 – 3

Activity
5, 10, 12, 9, 11, 7, 4, 6, 3, 8
12, 8, 7, 9, 6, 8, 13, 11, 10, 5

3:1

(1) **a** quarter past 6 or 15 minutes past 6 **b** quarter to 11 or 15 minutes to 11 (2) **a** 12, 15, 18, 21 **b** 16, 20, 24, 28 **c** 28, 35, 42, 49
(3) **a** 2, 2 **b** 3, 3 (4) $10 (5) **a** 10 **b** 10 **c** 9 **d** 1
(6) **a** jug **b** book (7) 4

3:2

(1) 20 (2) 17 (3) 5 (4) 2 (5) 27 (6) 6 (7) 12 (8) 2 (9) 6
(10) 26c (11) **a** 6 **b** 3 **c** 9 **d** 9 (12) 5 groups of 5 = 25
(13) **a** 26 **b** 39 (14) 7, 15, 17 (15) 14, 14, 12
(16) **a** 4 **b** 6 **c** 3
(17) **a** 550 will be circled. 500 will be underlined.
b 462 will be circled. 406 will be underlined.

Activity
5, 13, 9, 6, 10, 8, 12, 7, 11, 14
7, 15, 11, 8, 12, 10, 14, 9, 13, 16

3:3

(1) 12 (2) 16 (3) 14 (4) **a** 59 **b** 56 **c** 58 **d** 56
(5) 10 (6) 3, 8 (7) **a** 3 **b** 7 (8) **a** 6 **b** 8 **c** 5 **d** 8
(9) **a** 10 **b** 25 **c** 16 **d** 12 (10) **a** 7 **b** 12

3:4

(1) yes (2) **a** 17 **b** 14 (3) **a** 20 **b** 25 **c** 40
(4) **a** A **b** B (5) 180 minutes

Challenge
Answers will vary, e.g. 20c + 20c + 10c.

Activity
6, 9, 12, 11, 7, 15, 10, 14, 8, 13
7, 10, 13, 12, 8, 16, 11, 15, 9, 14

4:1

(1) 20 (2) 11 (3) 2 (4) 2 (5) 25 (6) 16 (7) 20
(8) 10 (9) 8 (10) 24c (11) **a** 2 **b** 2 **c** 4 **d** 4
(12) **a** oval b square (13) no (14) **a** 8, 10, 12, 14
b 70, 60, 50, 40 **c** 108, 110, 112 **d** 700, 600, 500
(15) 12, 12

4:2

(1) 22 (2) 30 (3) 24 (4) 7 (5) 32 (6) 18 (7) 28
(8) 4 (9) 4 (10) 31c (11) 180 (12) 8 (13) 12
(14) **a** 24, 30, 36, 42 **b** 40, 50, 60, 70 **c** 36, 45, 54, 63
(15) **a** 18 **b** 19 **c** 25 **d** 40 (16) yes (17) 12
(18) **a** 60 **b** 100 **c** 80

Activity
a S **b** N **c** N **d** S **e** A **f** S

4:3

(1) 20 (2) 20 (3) 19 (4) 60, 50, 40 (5) 653
(6) Answers will vary.
It has 6 faces, 12 edges and 8 corners. Its cross-section is a square. It can stack and slide.
(7) **a** 63, 53, 43, 33 **b** 680, 675, 670
c 559, 569, 579 (8) 2, 4, 6, 8 or 0 (9) impossible
(10) **a** 29 **b** 94 **c** 98 **d** 79
(11) **a** 37 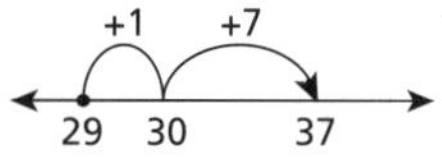

b 71 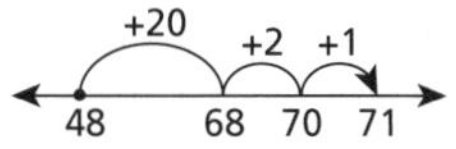

(12) **a** 4 **b** 9

4:4

(1) 350, 400, 450, 500 (2) 50 (3) 8 (4) 52 (5) 6, 3
(6) SMILE AT ME (7) **a** 80 **b** 80 (8) 14

Challenge
17, Answers will vary, e.g. 3 + 3 = 6, so I subtracted 3 from 23 to bridge to 20, then I subtracted 3 more from 20 to get 17.

Activity
8, 16, 12, 9, 13, 11, 15, 17, 10, 14
10, 18, 14, 11, 15, 13, 17, 19, 16, 13

5:1

(1) 30 (2) 15 (3) 9 (4) 5 (5) 46 (6) 30 (7) 6 (8) 6 (9) 2
(10) 24c (11) unlikely (12) 234 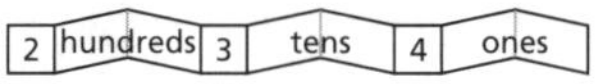

(13) two hundred and thirty-four (14) **a** 1 **b** 2
(15) 12, 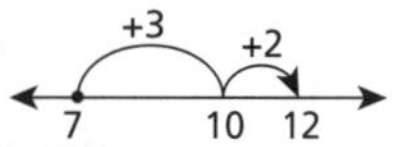

5:2

(1) 43 (2) 34 (3) 6 (4) 7 (5) 74 (6) 16 (7) 9 (8) 8
(9) 7 (10) 15c (11) no (12) 391, 535, 902 (13) 604

14 518 15 autumn 16 **a** 12, 15, 18, 21 **b** 46, 36, 26, 16 **c** 142, 132 **d** 445, 450 17 18 11th

Activity

13, 8, 16, 12, 14, 9, 17, 15, 11, 18

14, 9, 17, 13, 15, 10, 18, 16, 12, 19

5:3

1 41 2 34 3 51 4 a cube, 12 edges

b cylinder, 2 curved edges 5 5 hundreds 5 tens 0 ones

6 **a** 8, 10, 12, 14 **b** 53, 63, 73, 83 **c** 135, 125, 115

7 unlikely 8 twenty 9 Answers will vary. A cube, prism or cylinder could be drawn.

5:4

1 **a** 71 **b** 92 2 30 3 86 4 78 5

6 three hundred and forty-two

7 **a** yes **b** 4

Challenge

Answers will vary, e.g a face.

Activity

Answers may vary. Least likely to most likely: **C** my mother, **E** a

father, **B** the principal, **A** a teacher, **F** a boy, **D** a human

6:1

1 50 2 16 3 4 4 3 5 39 6 30 7 3 8 6

9 2 10 14c 11 4 hundreds, 9 tens, 3 ones

12 76, 129, 212, 236, 318

13

14

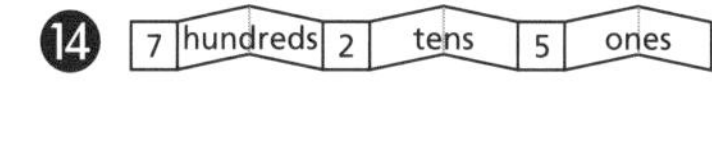

15 11, 13, 15, 17, 19 16 fifty

17 nine hundred and three 18 12th

6:2

1 36 2 22 3 21 4 23 5 38 6 40 7 3 8 2

9 46 10 21c 11 9 hundreds, 4 tens, 2 ones 12 30

13 758, 857, 875 14 359 15

16 **a** yes **b** yes **c** no

Activity

(5) triangle (6) square (7) rectangle (11) quadrilaterals (12) pentagons (13) hexagons (14) circle (15) oval (18) side, corner

6:3

1 48 2 71c 3 88 4 4 hundreds, 5 tens, 2 ones

5 **A** and **D** 6 **a** 43 (+4, +3; 36, 40, 43) **b** 74

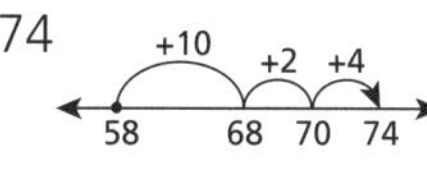

7

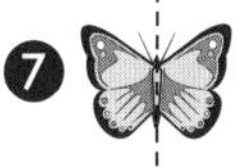

8 **a** 13 **b** 4 **c** 13 **d** 9 9 $140

6:4

1

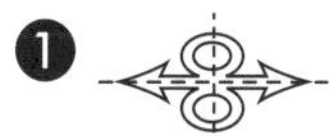

2 30 3 20 4 32 5 6 6 3

7 5 1 tens 8 ones

Challenge

Answers will vary, e.g. A or E.

Activity

a 7 + 6 = 13, 6 + 7 = 13, 13 – 7 = 6, 13 – 6 = 7

b 8 + 5 = 13, 5 + 8 = 13, 13 – 8 = 5, 13 – 5 = 8

c 4 + 8 = 12, 8 + 4 = 12, 12 – 8 = 4, 12 – 4 = 8

7:1

1 63 2 25 3 4 4 2 5 46 6 20 7 10 8 10

9 30 10 12c 11 6 hundreds, 3 tens, 5 ones

12 13 $100, $20, $5 14 A

15 The dots will be joined to make a pentagon. This is a regular pentagon. 16 circle

7:2

1 43 2 49 3 6 4 12 5 49 6 15 7 5 8 20

9 40 10 42c 11 5 hundreds, 8 tens, 6 ones

12 **a** 8 **b** 5 13 80 14 $180 15

16 **a** 15 **b** 7 **c** 15 **d** 8 17 20

18 462 19 3

Activity

12, 0, 10, 16, 4, 14, 8, 20, 6, 18

0, 0, 0, 0, 0, 0, 0, 0, 0, 0

7:3

1 49 2 42c 3 **a** rectangular prism **b** 6 **c** 8 **d** 12 **e** a rectangle 4 14, 14 – 5 = 9, 14 – 9 = 5

5 **a** 20 **b** 20 **c** 50 **d** 50 6 **a** D **b** B 7 555

7:4

1  **2** 53 tens and 1 one **3** 4 or 5 **4** 34

Challenge

Answers will vary. One answer could be:
This is called a triangular prism. It has 6 corners, 5 faces, 9 edges and 2 triangular bases. All other faces are rectangles. It can stack and slide.
A Toblerone chocolate bar is the shape of a triangular prism.

Activity

10, 20, 30, 40, 50, 60, 70, 80, 90, 100

8:1

1 30 **2** 50 **3** 16 **4** 20 **5** 51 **6** 50 **7** 10
8 60 **9** 70 **10** 31c **11** 18 **12** regular

13

	3	7	4	5	2	10	6	9
× 5	15	35	20	25	10	50	30	45

14 the width of a door **15** 2 rows of 5 = 10 **16** 14

17 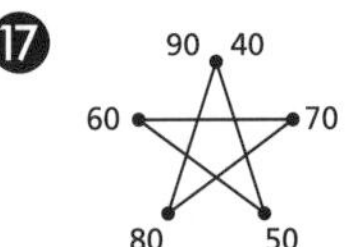

8:2

1 70 **2** 50 **3** 8 **4** 16 **5** 97 **6** 40 **7** 30
8 50 **9** 100 **10** 71c **11** The dots will be joined to make a hexagon. This is an irregular hexagon.

12 90 metres **13** 63 + 14 = 77

14 Two parallel lines will be drawn.

15 one hundred and twenty-eight

16 **a** 709 **b** 998

Activity

30, 0, 25, 40, 10, 35, 20, 50, 15, 45
12, 0, 10, 16, 4, 14, 8, 20, 6, 18

8:3

1 67 **2** 24c

3 **a** 28 −2 −1 −20 28 30 31 51 **b** 63 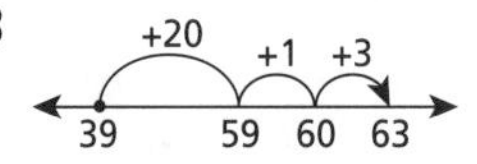

4 **a** 9 **b** 19 **c** 29 **d** 39 **e** 43 **f** 53 **g** 63 **h** 73

5 **a** F **b** C **c** A **d** B **6** 731

8:4

1 **a** 10 **b** 20 **2** **a** 26 **b** 44 **c** 23 **d** 36

3 **a** 0 **b** 7 **4** 3 **5** Estimates will vary. 5 groups of 10 will be circled, plus 2 stars. 52

Challenge

77. Answers will vary, e.g. I used the split strategy. I added the tens first. 4 tens and 3 tens makes 7 tens. Then I added the ones.

2 and 5 makes 7. 7 tens and 7 ones makes 77.

Activity

5, 10, 15, 20, 25, 30, 35, 40, 45, 50

	2	4	6	8	10
× 5	10	20	30	40	50

	1	2	3	4	5
× 10	10	20	30	40	50

9:1

1 38 **2** 48 **3** 58 **4** 68 **5** 50 **6** 10 **7** 12
8 80 **9** 40 **10** 50c **11** 5, 5, 5 **12** 7

13 Answers will vary. **14** 15 past 12 or quarter past 12

15 60 **16** nine thirty or half past 9

17 5 + 6 = 11 **18** 

9:2

1 63 **2** 73 **3** 83 **4** 93 **5** 79 **6** 14 **7** 35
8 30 **9** 90 **10** 59c **11** 18 − 13 = 5 so 13 + 5 = 18

The difference between 13 and 18 is 5. **12** 4 metres

13 **a** 8 : 15 **b** 9 : 30

14 **a** 30 **b** 60 **15** **a** 12 **b** 18

16 **a** 95 **b** 15 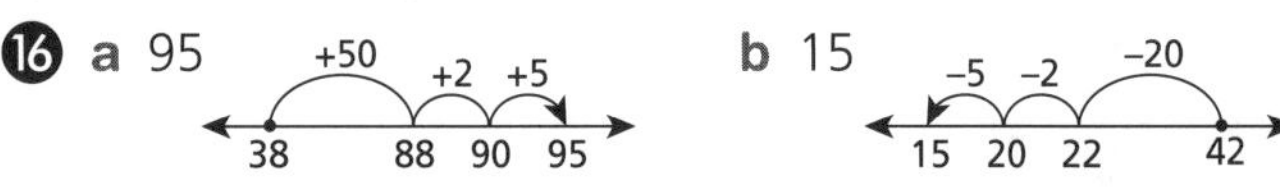

Activity

0, 3, 5, 7, 10, 8, 2, 6, 4, 9
0, 30, 50, 70, 100, 80, 20, 60, 40, 90

9:3

1 88 **2** 42c **3** $10

4 45 past 12 or quarter to 1

5 5, 5, 5

6 **a** 85 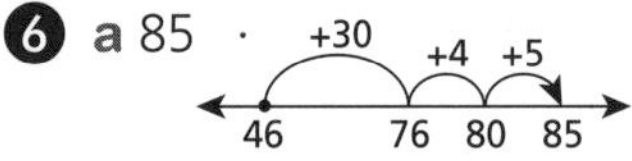

b 15 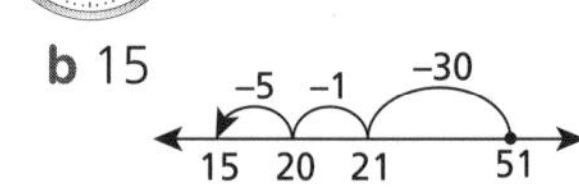

 • *AUSTRALIAN SIGNPOST MATHS 3 MENTALS* • ISBN 978 0 6557 0883 4

7 3 fives or 15 **8** 3 groups of <u>4</u> = <u>12</u>, 12

9 8 **10** 14 (days)

9:4

1 76 – 27 = <u>49</u> so 27 + <u>49</u> = 76 The difference between 27 and 76 is <u>49</u>. **2** 28 **3** one quarter **4** 8 + 8

5 6 **6** **a** 27 **b** 40 **c** 32 **7** **a** 8 **b** 13

Challenge

16 + 24 = 40, 24 + 16 = 40, 40 – 16 = 24, 40 – 24 = 16

Activity

	29	50	64	37	82	48	73	96	101	129
–10	19	40	54	27	72	38	63	86	91	119
–9	20	41	55	28	73	39	64	87	92	120

	31	78	80	65	52	66	27	94	113	139
–9	22	69	71	56	43	57	18	85	104	130

10:1

1 35 **2** 60 **3** 6 **4** 4 **5** 42 **6** 15 **7** 5 **8** 20

9 40 **10** 42c **11** 6, 6, 6 **12** C (m)

13 **a** 6:00 **b** 10:00 **14** 30

15 80c **16** B **17** 34

+4 +4

26 30 34

10:2

1 51 **2** 49 **3** 8 **4** 16 **5** 69 **6** 20 **7** 40

8 90 **9** 100 **10** 33c **11** yes (most year 3 students will be over 1 m in height) **12** 14, 14, 14 **13** 10 **14** **a** 6

b 16 **c** 26 **d** 36 **e** 43 **f** 53 **g** 63 **h** 73 **15** 10

16 <u>34</u> + <u>8</u> = <u>42</u> **17** half past 2 **18** 1 m

19 1 litre bottle

Activity

a 7 – 4 = <u>3</u>, 4 + <u>3</u> = 7 The difference is <u>3</u>.

b 10 – 8 = <u>2</u>, 8 + <u>2</u> = 10 The difference is <u>2</u>.

c 13 – 5 = <u>8</u>, <u>8</u> + 5 = 13 The difference is <u>8</u>.

d 28 – 12 = <u>16</u>, 12 + <u>16</u> = 28 The difference is <u>16</u>.

e 34 – 12 = <u>22</u>, 12 + <u>22</u> = 34 The difference is <u>22</u>.

f 44 – 36 = <u>8</u>, 36 + <u>8</u> = 44 The difference is <u>8</u>.

10:3

1 88 **2** 20c **3** 98 **4** 60 **5** yes

6 **a** 7:30 **b** 5:45

7 <u>45</u> past <u>8</u> or <u>a quarter</u> to <u>9</u>

8 9, 9, 9

9 **a** 82

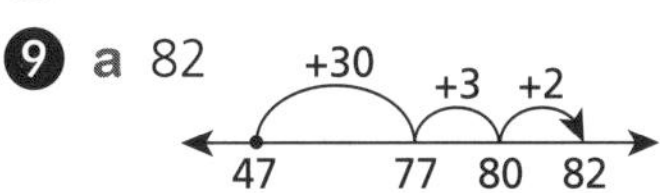

b 28

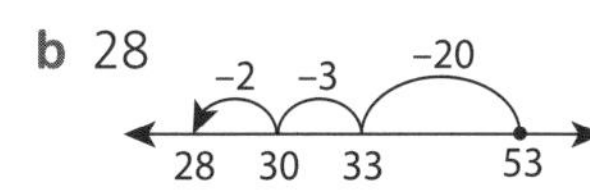

10 20, 5 groups of <u>4</u> = <u>20</u>, 20

10:4

1 $56 **2** 32 **3** The $2 and 20-cent coins will be coloured. **4** $7.50 **5** $2.85

Challenge

Answers will vary. Some examples are: $1 + $1 + $1, $1 + $1 + 50c + 50c, $2 + 50c + 20c + 20c + 10c, $1 + $1 + 50c + 20c + 20c + 10c, $2 + $1, $1 + $1 + 50c + 20c + 10c + 10c + 10c.

Activity

a 6, 8, 10, 12, 14 **b** 15, 20, 25, 30, 35

c 20, 30, 40, 50, 60, 70, 80, 90

11:1

1 18 **2** 28 **3** 38 **4** 48 **5** 80 **6** 20 **7** 10

8 20 **9** 50 **10** 50c **11** 6 metres **12** 50c coin **13** 60c

14 a 7 b 33 **15** 55, 61, 59, 64, 58 and 56 will be circled.

16 <u>5</u> hundreds, <u>6</u> tens, <u>3</u> ones **17** 6 **18** **a** 5 **b** 2

11:2

1 54 **2** 44 **3** 34 **4** 24 **5** 99 **6** 12 **7** 16

8 40 **9** 45 **10** 32c **11** yes **12** **a** 30 **b** $1.50

13 85c **14** 412, 392, 406, 351 **15** $110

16 <u>4</u> hundreds, <u>8</u> tens, <u>2</u> ones **17** 375, 385, 395

18 **a** 265 **b** 534

Activity

a 40 **b** 30 **c** 40 **d** 20 **e** 90

11:3

1 98 **2** 62c **3** 92 **4** yes **5** 824, 809, 750, 751, 793, 783 and 849 will be circled. **6** 9 hundreds, 0 tens, 4 ones

7 **a** One rectangle will be coloured out of 2.

b Two rectangles will be coloured out of 4.

c Four rectangles will be coloured out of 8.

8 **a** 3 out of 8 circles will be coloured.

b 5 out of 8 rectangles will be coloured.

c $\frac{3}{8}$ **9** **a** 8, 10, 12, 14, 16 **b** 40, 50, 60, 70, 80

c 53, 63, 73, 83, 93

11:4

1 40 **2** **a** 21 **b** 42 **3** a star **4** 3 **5** 21
6 century

Challenge

Answers will vary, e.g. a bath, swimming pool or sink.

Activity

a 8 **b** 20

12:1

1 7 **2** 6 **3** 6 **4** 3 **5** 80 **6** 25 **7** 10 **8** 50
9 15 **10** 80c **11** 9 hundred, 3 tens, 5 ones
12 74, 68, 71, 65, 66, 69, 67 and 73 will be circled.
13 centimetres, metres **14** **a** A **b** C
15 **a** One rectangle will be coloured. **b** One rectangle will be coloured. **c** One rectangle will be coloured. **d** One rectangle will be coloured.
16 4 hundred, 7 tens, 3 ones

12:2

1 9 **2** 8 **3** 50 **4** 90 **5** 78 **6** 16 **7** 40 **8** 80
9 30 **10** $6 **11** 100 **12** true, Answers will vary.
13 **a** C **b** A **14** 4 out of 9, $\frac{4}{9}$ **15** **a** 328, 349, 250 and 261 will be circled. **16** 7 teddy bears will be circled $\frac{7}{10}$.

Activity

a 300 **b** 700 **c** 300 **d** 800 **e** 800 **f** 500 **g** 100

12:3

1 69 **2** $15 **3** 78 **4** 5 out of 10, $\frac{5}{10}$ **5** yes
6 100 **7** 21 cm **8** 937, 851, 949 and 850
9 B and C **10** **a** 8, 10, 12, 14 **b** 12, 15, 18, 21
c 20, 25, 30, 35 **d** 40, 50, 60 **11** 5 groups of 3
12 6, 9, 12

12:4

1 99 cm **2** half **3** 24 **4** **a** 6 **b** 3 **5** 3 **6** C
7 4 **8** **a** 25 **b** 30

Challenge

Any number from 650 to 749 could be written.

Activity

a 5 out of 8 or $\frac{5}{8}$ **b** 2 out of 5 or $\frac{2}{5}$ **c** 5 out of 6 or $\frac{5}{6}$
d 6 out of 7 or $\frac{6}{7}$

13:1

1 80 **2** 30 **3** 6 **4** 8 **5** 70 **6** 15 **7** 20 **8** 30
9 40 **10** 10c **11** 1 out of 2 equal parts
12 **a** circle **b** rectangle **13** cm **14** yes **15** 4 out of 6
16 **a** 4 **b** 5 **c** 3 **d** 8 **e** 1 **f** 6

13:2

1 59 **2** 5 **3** 10 **4** 12 **5** 89 **6** 25 **7** 30
8 50 **9** 60 **10** 40c **11** 1 m 65 cm **12** 24 cm
13 3 out of 7 **14** false **15** **A** 4 cm **B** 3 cm **C** 5 cm
16 30

Activity

a 39
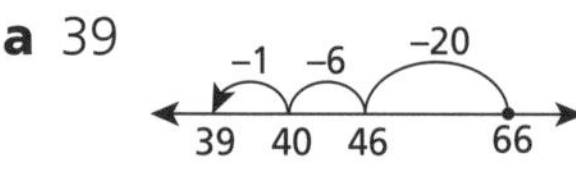

b 37
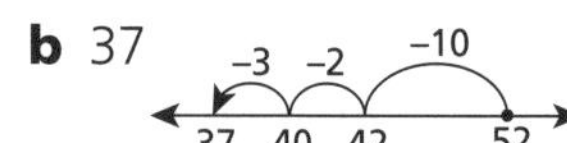

c 35
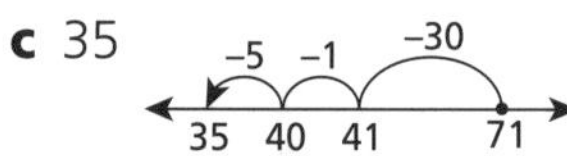

13:3

1 33 **2** 47 **3** 50 **4** 28 cm **5** 4 m 27 cm
6 one half **7** 5 cm **8** cm **9** 6 + 8 = 14 **10** yes
11 82 (+30, +3, +2: 47, 77, 80, 82) **12** 941

13:4

1 300
2 **a** **b**
c 2 seals will be coloured
3 4 **4** **a**  **b** 2 **c** $\frac{1}{2}$ **5** **a** 3 out of 4
b 4 quarters or 1 **6** **a** 15 – 6 = 9 **b** 15 – 9 = 6

Challenge

15 (– 5, – 3, – 20: 15, 20, 23, 43)

Activity

+	5	10	4	0	2	6	1	3
2	7	12	6	2	4	8	3	5
3	8	13	7	3	5	9	4	6
4	9	14	8	4	6	10	5	7
5	10	15	9	5	7	11	6	8
10	15	20	14	10	12	16	11	13

 • *AUSTRALIAN SIGNPOST MATHS 3 MENTALS* • ISBN 978 0 6557 0883 4

14:1

❶ 20 ❷ 12 ❸ 17 ❹ 8 ❺ 70 ❻ 11 ❼ 4 ❽ 8
❾ 3 ❿ 10c ⓫ m ⓬
⓭ 77 (+10, +1, +7; 59, 69, 70, 77)

⓮ a 500 b 26 c 53 d 28 ⓯ 32 – 16 = 16, 16

⓰ a 16, 18, 20, 22 b 25, 30, 35, 40 c 66, 76, 86, 96

14:2

❶ 84 ❷ 5 ❸ 83 ❹ 24 ❺ 75 ❻ 15 ❼ 14
❽ 90 ❾ 8 ❿ 33c ⓫ D ⓬ 6 m 93 cm
⓭ 135 cm ⓮

⓯ 18 (– 2, – 4, – 10; 18, 20, 24, 34) ⓰ a 57 b 46 ⓱ 24

Activity

5, 10, 15, 20, 25, 30, 35, 40, 45, 50

	2	4	6	8	10
× 5	10	20	30	40	50

	2	4	6	8	10
× 10	20	40	60	80	100

14:3

❶ 94 ❷ $77 ❸ a 64 b 63 c 94 d 64 e 37 f 36

❹ a Brown St b Jones St c Bond St d Smith St
e Bond St, Smith St and Brown St ❺ 37 (– 3, – 1, – 20; 37, 40, 41, 61)

14:4

❶ 700 ❷ 1 m 50 cm ❸ 20 ❹ C

❺ a True b True ❻ D

Challenge

64, Answers will vary, e.g. I used the compensation strategy. I know that 40 + 25 = 40 + 20 + 5 which equals 65. 39 is one less than 40 so the answer is one less than 65. The answer is 64.

Activity

a 14 cm b 12 cm c 22 cm

15:1

❶ 33 ❷ 12 ❸ 50 ❹ 18 ❺ 48 ❻ 16 ❼ 25
❽ 70 ❾ 90 ❿ 22c ⓫ 1357 ⓬ yes ⓭ metres

⓮ a blue b orange c 12 d 10

⓯ a 700 b 700 c 800

15:2

❶ 55 ❷ 52 ❸ 800 ❹ 57 ❺ 69 ❻ 18 ❼ 45
❽ 80 ❾ 100 ❿ 31c ⓫ no (although the largest recorded foot length for a person is about 47 cm)

⓬ 9531 ⓭ 472 cm ⓮ a Greg b Sue c 7 ⓯ 9

Activity

10 = 1 + 9 = 8 + 2 = 3 + 7 = 5 + 5 = 7 + 3 = 2 + 8 = 6 + 4 = 4 + 6

13 = 3 + 10 = 6 + 7 = 4 + 9 = 7 + 6 = 8 + 5 = 9 + 4 = 5 + 8 = 10 + 3

15:3

❶ $97 ❷ $84 ❸ 1357 ❹ 4 ❺ 5423

❻ a 53 b 35

❼ a 73 (+ 20, + 4, + 3; 46, 66, 70, 73) b 46 (– 4, – 3, – 20; 46, 50, 53, 73)

❽ a 65 b 67 ❾ even ❿ a 5329 b 9517

15:4

❶ 40 ❷ Possible answers are 4204, 4206, 4208, 4210 or 4212.

❸ 60 ❹ a 6 b 3 ❺ 32 ❻ 9, 8 and 7 ❼ yes ❽ 6031

Challenge

55, Answers will vary, e.g. I used the compensation strategy. I know that 84 – 30 = 54 because 8 tens minus 3 tens is 5 tens. 29 is one less than 30 so I need to take away one less so the answer is 55.

Activity

14, 17, 11, 19, 12, 15, 18, 20, 10, 16
30, 60, 41, 54, 62, 85, 44, 67, 73, 96

16:1

❶ 400 ❷ 500 ❸ 56 ❹ 55 ❺ 88 ❻ 3 ❼ 8
❽ 14 ❾ 30 ❿ 47c ⓫ a dog b mouse c 13 d 7
⓬ 6572

⓭ 10 : 22 ⓮ 5911 ⓯ 4

⓰ a 9th b 7th

16:2

❶ 900 ❷ 300 ❸ 91 ❹ 14 ❺ 89 ❻ 3 ❼ 9
❽ 18 ❾ 40 ❿ 8c ⓫ $\frac{4}{8}$ or $\frac{1}{2}$

 ISBN 978 0 6557 0883 4

⓬ four thousand, five hundred and one

⓭ **a** Leah **b** Anna **c** 7 **d** 26

⓮ 3 thousands, 5 hundreds, 2 tens, 3 ones ⓯ even

Activity

a 60 **b** 30 **c** 15 **d** 45 **e** 50 **f** 10

16:3

❶ 700 ❷ 900 ❸ 900 ❹ 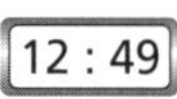

❺ 1357

❻ 3 out of 8 squares will be coloured.

❼ 7020 will be circled.

❽

Kangaroos							
Emus							
Rabbits							

❾ **a** 54 **b** 8

16:4

❶ a 10:51 **b** 11:58 ❷ 11 ❸ 4 blocks will be circled.

❹ 28 ❺ 30 ❻ 36 ❼ 73 ❽ **a** one quarter **b** one half **c** three quarters

Challenge

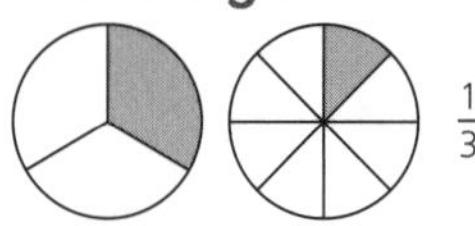

$\frac{1}{3}$

Activity

a IV **b** V **c** IX **d** X **e** XI **f** XII **g** 9 **h** 11 **i** 6

17:1

❶ 6 ❷ 3 ❸ 10 ❹ 5 ❺ 88 ❻ 3 ❼ 8 ❽ 14

❾ 30 ❿ 47c ⓫ kg ⓬ less ⓭ **a** 6 **b** 9

⓮ One fish will be coloured. ⓯ 6:53

⓰ 4 kg will be circled. ⓱ 645, 745, 845

17:2

❶ 15 ❷ 25 ❸ 20 ❹ 50 ❺ 88 ❻ 3 ❼ 2 ❽ 6

❾ 52 ❿ 13c ⓫ true ⓬ 3 past 9 ⓭ 3 kg ⓮ 4093

⓯ **a** 9 to 7 **b** 21 to 11 ⓰ 2 out of 4, $\frac{2}{4}$

⓱ 12, 15, 18, 21, 24

Activity

3, 6, 9, 12, 15, 18, 21, 24, 27, 30

17:3

❶ **a**  **b** ❷ **a** less **b** more

❸ 4:02 ❹ 4:57 ❺ 1537 ❻ m ❼ 3 m 35 cm

❽ fourteen, forty, forty-four ❾ **a** 57, 47, 37, 27, 17 **b** 24, 22, 20, 18, 16

17:4

❶ 3 ❷ 4 blocks will be circled. ❸ true

❹ **a** 5 squares will be coloured red and 1 square will be coloured blue. **b** 6 sixths, 1 whole or $\frac{6}{6}$

❺ 24 ❻ 3:33 ❼ + ❽ no

Challenge

Answers will vary, e.g a pencil case, pair of shoes or hat.

Activity

9, 12, 15, 18, 21

12, 16, 20, 24, 28

10, 15, 20, 25, 30, 35, 40, 45

18:1

❶ 4 ❷ 8 ❸ 6 ❹ 4 ❺ 65 ❻ 10 ❼ 4 ❽ 12

❾ 35 ❿ 2c ⓫ yes ⓬ 6, 6 ⓭ 12 ⓮ kilograms

⓯ 20, 20 ⓰ Answers will vary. ⓱ C ⓲ 6

⓳ 3 groups of 5 balls will be drawn, 15

18:2

❶ 80 ❷ 8 ❸ 16 ❹ 8 ❺ 77 ❻ 12 ❼ 16 ❽ 40

❾ 45 ❿ 21c ⓫ 8 kg ⓬ 16 ⓭ **a** hexagon **b** rhombus ⓮ **a** 64 **b** 86 **c** 100 **d** 42

⓯ square, rhombus ⓰ no. Some birds can't fly, including penguins and emus.

⓱ **a** 45, 35, 25, 15, 5 **b** 12, 15, 18, 21, 24

Activity

a 5 **b** Answers will vary.

18:3

❶ 24 ❷ yes ❸ parallelogram ❹ **a** yes **b** yes **c** yes

❺ 4 rows of 6 = 24, 4 x 6 = 24 ❻ yes

❼ 1, 2, 3, 4, 5, 6 ❽ yes ❾ $\frac{6}{8}$ ❿ 8, 8, 16

⓫ 4 rows of 5 balls will be drawn. 20

18:4

1 **a** 60c **b** 280c or $2.80 **2** 5 **3** 25

4 **a** 1, 9, 7 **b** 3, 9, 7 **5** +

Challenge

Answers will vary, e.g. a square could be drawn. It has 4 equal
sides and 4 vertices. It is a quadrilateral. It has 2 sets of parallel sides.

Activity

0, 10, 16, 4, 12, 18, 8, 20, 6, 8
4, 8, 12, 16, 20
4, 8, 12, 16, 20

19:1

1 8 **2** 12 **3** 40 **4** 20 **5** 39 **6** 18 **7** 9 **8** 55

9 55 **10** $45 **11** 44, 48, 52, 56, 60 **12** 24 **13** ×

14 **a** 20, 25, 30 **b** 18, 16, 14 **c** 14, 24, 34

d 50, 40, 30 **15** **a** 3 **b** 4 **16** C

19:2

1 24 **2** 32 **3** 36 **4** 24 **5** 94 **6** 27 **7** 59

8 13 **9** 89 **10** $10 **11** **a** true **b** true **12** 20

13 **a** 16, 12, 8, 4 **b** 46, 36, 26, 16 **c** 34, 24, 14, 4

d 77, 79, 81, 83 **14** yes

15 The angle will be copied.

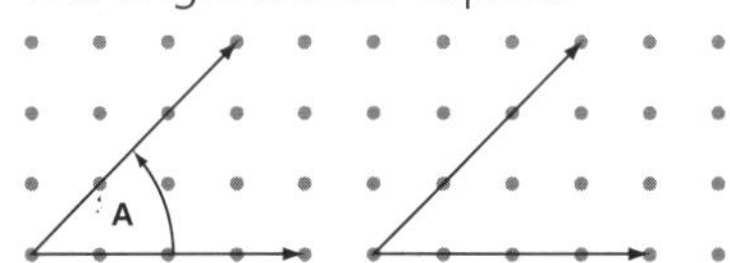

Activity

a likely or unlikely (Answers may vary.) **b** certain
c unlikely **d** impossible **e** even chance

19:3

1 30 **2** 19 **3** 21 **4** **a** 16, 32, 64 **b** 24, 48, 96

c 32, 64, 128 **d** 200, 400, 800 **5** **a** **C** **b** **A**

6 45, 50, 55, 60 **7** right angle **8** yes

9 **a** 40 **b** 20 **c** 28 **d** yes

19:4

1 **a** triangle **b** rectangle **2** circle **3** **a** 20 **b** $1\frac{2}{4}$

c Answers may vary, e.g. $2\frac{2}{4}$, $2\frac{1}{2}$ or 10 quarters **4** 26

Challenge

Answers will vary. There is an equal chance of getting an A or B.
It is impossible to spin a C. It is possible to get an A or a B.

Activity

24, 3, 15, 18, 6, 27, 21, 9, 30, 12
32, 4, 20, 24, 8, 36, 28, 12, 40, 16

20:1

1 12 **2** 16 **3** 6 **4** 15 **5** 43 **6** 8 **7** 12 **8** 9

9 6 **10** $10 **11** yes **12** **C**, **A**, **B** **13** no **14** yes

15 12 **16** 12, 9, 6, 3 **17** 9 **18** **a** **b** 4 **c** 4

19 $25

20:2

1 28 **2** 36 **3** 18 **4** 21 **5** 97 **6** 18 **7** 15

8 18 **9** 29 **10** $20 **11**

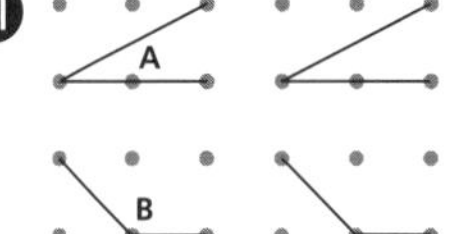

12 **a** 1000, 10 000 **b** 2000, 20 000 **c** 3000, 30 000
d 4000, 40 000

13 parallel lines **14** no

15 12, 15, 18, 21, 24 **16** yes **17** $28

Activity

(3) parallel lines (4) perpendicular lines (8) diamond or rhombus (9) trapezium (10) parallelogram
(11) quadrilaterals (12) pentagons (13) hexagons

20:3

1 25 **2** 21 **3** 28 **4** yes

5 **a** 9, 27 **b** 6, 18 **c** 9, 27 **d** 12, 36

6 **A**, **C**, **B**, **D** **7** **B** **8** no **9** 21

10 **a** none or 0 **b** 6 **c** 8

20:4

1 7 (1+15, 2+14, 3+13, 4+12, 5+11, 6+10, 7+9)

2 a 12 b 14 **3** **a** **F** **b** **B** and **D** **4** $2

5 **a** 30c **b** 105c or $1.05

Challenge

Answers may vary.
It has 4 sides, 4 corners and 4 angles.
Its opposite sides are parallel.

Activity

1 **a** unlikely **b** likely **c** impossible **d** certain **2** less

21:1

1 0 **2** 12 **3** 8 **4** 16 **5** 77 **6** 3 **7** 0 **8** 11
9 44 **10** $33 **11** C **12** 25 **13** angle
14 **a** rhombus **b** hexagon **c** 3 **d** 7

21:2

1 21 **2** 27 **3** 24 **4** 32 **5** 89 **6** 36 **7** 27
8 4 **9** 12 **10** $50 **11** **a** 18 **b** 16
12 2 parts will be coloured. **13** 40 **14** 500 **15** no
16 2 **17** 44 and 46
18 28

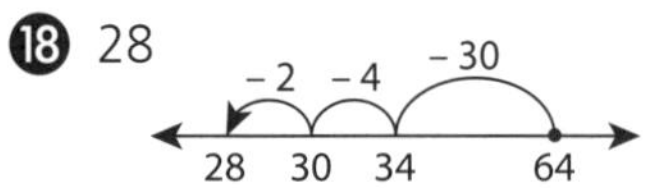

19 382

Activity

3, 24, 21, 27, 12, 18, 9, 30, 15, 6
4, 32, 28, 36, 16, 24, 12, 40, 20, 8

21:3

1 89 **2** 99 **3** 3 parts will be coloured **4** 3 out of 4
5 **a** A **b** C **6** **a** Red **b** 8 **c** 60 **7** **a** 8 **b** 12 **c** 10

21:4

1 6 boats will be coloured blue, 4 will be coloured red.
2 1000 mL or 1 L **3** 3 **4** 61

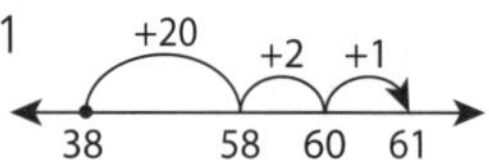

5 **a** 68 **b** 98

Challenge

Answers will vary.

Activity

4, 7, 1, 3, 6, 2, 5, 0
5, 8, 2, 4, 7, 3, 6, 1

22:1

1 9 **2** 18 **3** 4 **4** 20 **5** 77 **6** 6 **7** 12 **8** 10
9 22 **10** $14 **11** angle **12** white
13 **a** 7 squares will be coloured.
b 5 squares will be coloured.
c $\frac{7}{8}$ **14** 5006 **15**
16 4 thousands + 5 hundreds

22:2

1 30 **2** 27 **3** 32 **4** 28 **5** 49 **6** 64 **7** 49
8 43 **9** 54 **10** $52 **11** **a** Jill **b** Kira **c** 7 **d** 5 **e** 27
12 , 9 **13** **a** 4 **b** 6

Activity

7, 11, 2, 5, 10, 8, 6, 4, 9, 3
8, 12, 3, 6, 11, 9, 7, 5, 10, 4,

22:3

1 30 **2** 19 **3** 21
4 5 thousands + 7 hundreds + 3 tens + 5 ones **5** 6031
6

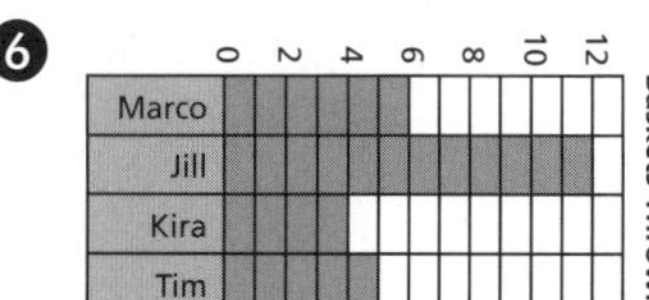

7 C **8** **a** 7483 **b** 8

22:4

1 **a** $3 **b** 85c **2** **a** 6 **b** 12
3 6 rectangles will be coloured.
4 3200 **5** $120 **6** $\frac{2}{3}$ **7** 16 **8** 5288

Challenge

Answers will vary.

Activity

3, 8, 4, 9, 6, 2, 10, 5, 11, 7
4, 9, 5, 10, 7, 3, 11, 6, 12, 8

23:1

1 20 **2** 6 **3** 18 **4** 15 **5** 79 **6** 12 **7** 9 **8** $15
9 $12 **10** $54 **11** 9
12 7 thousands + 2 hundreds + 9 tens + 4 ones **13** **A**
14 $\frac{5}{8}$ **15** 30 **16** 7

23:2

1 34 **2** 42 **3** 67 **4** 83 **5** 77 **6** 7 **7** 55 **8** 27
9 38 **10** $42 **11** $\frac{3}{4}$
12 9 thousands + 7 hundreds + 2 tens + 9 ones
13 62 **14** March, April, May **15** 60 **16** 14 **17** 13.4.25
18 3278 **19** yes **20** **a** 1 4 hundreds 7 tens 9 ones
b 1 4 7 tens 9 ones

Activity

January, February, March, April, May, June, July, August, September, October, November, December

23:3

1 **a** $\frac{3}{8}$ **b** 4 rectangles will be coloured. $\frac{4}{8}$ **c** $\frac{4}{8}$ will be ticked. **2** **a** October **b** 5 **c** Friday **d** 31 **e** 7th October or October 7 **f** 27th of October or October 27

3 5 **4** 8 thousands + 2 ones **5** **a** 12, 15, 18, 21, 24 **b** 20, 25, 30, 35, 40

23:4

1 **a** 8 **b** 10 **2** **a** 120c or $1.20 **b** 300c or $3

3 3390 **4** Answers will vary, e.g. $\frac{9}{10}$ or $\frac{99}{100}$. The higher the number at the base of the fraction, the smaller the parts of a whole will be. The higher the number at the top, the larger the fraction will be.

5 9 **6** **a** 38 **b** 32

Challenge

Answers will vary.

Activity

14, 22, 26, 19, 37, 57, 15, 29, 32, 43

8, 16, 20, 13, 31, 51, 9, 23, 26, 37

24:1

1 9 **2** 0 **3** 20 **4** 40 **5** 10 **6** 15c **7** $6 **8** 18

9 **a** $\frac{3}{4}$ **b** Answers may vary, e.g. $\frac{2}{4}$ or $\frac{1}{2}$.

10 **a** R **b** J **c** D **d** L **e** B **11** 4931

12

	3	7	4	5	2	10	6	9
× 2	6	14	8	10	4	20	12	18

13 14

24:2

1 12 **2** 0 **3** 10 **4** 20 **5** 12 **6** 15c **7** $2

8 8 **9** December

10 **a** 5 **b** Tuesday **c** Sunday (7th of June)

11 **a** B **b** A, B, C **12** 3, 3

Activity

a 3 **b** 4 **c** 6 **d** 9

24:3

1 **a** 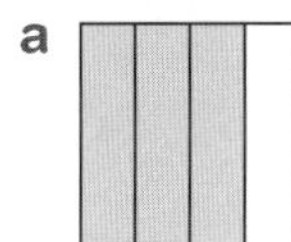 **b** 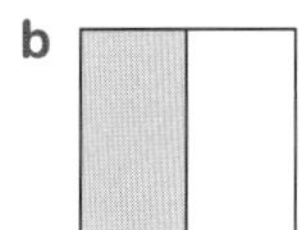 **c** $\frac{3}{4}$

2

3 6 **4** **a** 12 **b** February **c** January **d** 11th **e** 5.5.2026 or 5.5.26

5

	3	7	4	5	2	10	6	9
× 4	12	28	16	20	8	40	24	36

6 **a** 48, 38, 28, 18 **b** 26, 28, 30, 32

24:4

1 2 bells will be coloured red. 4 bells will be coloured blue. $\frac{6}{8}$ or $\frac{3}{4}$

2 **a** one fifth b one twentieth **3** 9089 **4** 72 **5** 60

6 **a** Sunday **b** Tuesday **7** **a** 35 **b** 70

8 **a** $16 **b** $127

Challenge

Answers will vary.

Activity

2, 6, 10, 4, 9, 1, 7, 5, 3, 8

13, 57, 42, 18, 56, 21, 39, 25, 44, 30

25:1

1 20 **2** 20 **3** 20 **4** 20 **5** 36 **6** 12 **7** 15

8 13 **9** **a** 30 **b** 30 **c** 31 **d** 31 **10** **a** 8, 8 **b** 6, 6

11 **a** 8 **b** 4 **c** 4 **12** grams **13** 30 g

14

	3	7	4	5	2	10	6	9
× 3	9	21	12	15	6	30	18	27

15 **a** 50, 45, 40, 35, 30 **b** 46, 48, 50, 52, 54 **c** 56, 46, 36, 26, 16

25:2

1 13 **2** 56 **3** 30 **4** 38 **5** 58 **6** 34 **7** 23

8 70 **9** 6, 6 **10** 7, 7

11 **a** Row 2 **b** Fruit **c** Soap **d** Row 1 **e** Lollies

Activity

a 2 rows of 3 = 6, How many 3s in 6? 2, 2 × 3 = 6 so 6 ÷ 3 = 2

b 3 rows of 5 = 15, How many 5s in 15? 3, 3 × 5 = 15 so 15 ÷ 3 = 5

c 3 groups of 3 = 9, How many 3s in 9? 3, 3 × 3 = 9 so 9 ÷ 3 = 3

25:3

❶ 8,8 ❷ 20, 20, 10 ❸ 40, 4, 10 ❹ grams

❺ 6, 3, 2 ❻ 67 mL

❼ 250 g ❽ 500 mL ❾ 85

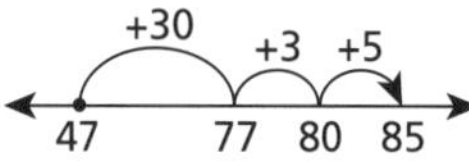

25:4

❶ **a** 4 **b** 3 ❷ 6 ❸ 9 ❹ **a** yes **b** 3

❺ 1000 g or 1 kg

Challenge

Answers will vary, e.g. pasta 350 g.

Activity

20, 31, 26, 19, 15, 40, 49, 64, 25, 23
21, 25, 44, 15, 37, 42, 50, 53, 28, 39

26:1

❶ 6 ❷ 6 ❸ 9 ❹ 9 ❺ 5 ❻ 5 ❼ 3 ❽ 3

❾

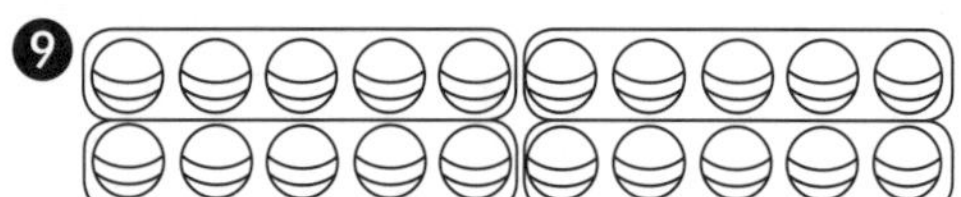

20, 20, 4, 4, 5, 5 ❿ 100 grams ⓫ 500 g ⓬ 6, 6

⓭ 3 ⓮ 27 mL ⓯ **a** 7 **b** 30 **c** 31 **d** 30 ⓰ 4396

26:2

❶ 4 ❷ 4 ❸ 4 ❹ 4 ❺ 9 ❻ 9 ❼ 7 ❽ 7

❾

24, 24, 6, 6, 4, 4

❿ **a** 9 **b** 18 **c** 9 ⓫ 500, 250

⓬ 750 mL ⓭ 33 (−7, −1, −20; 33, 40, 41, 61)

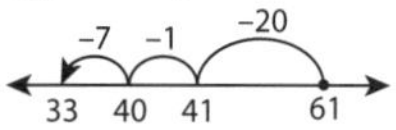

⓮ millilitres ⓯ 4291 ⓰ 2

Activity

(9) even numbers (10) odd numbers
(11) ordinal numbers (12) digits (18) number line
(19) number sentence (20) numeral expander
(21) abacus

26:3

❶ 39 mL ❷ 2 ❸ 28 kilograms

❹ **a** 20, 5, 5 **b** 18, 6, 6

❺ **a** Pa **b** Nanna **c** Alan **d** Diane **e** Lyn

❻ **a** 6, 6 **b** 9, 9 **c** 8, 8 **d** 6, 6

26:4

❶ **a** 12 **b** 9 ❷ **a** 51 **d** 32 ❸ 114 ❹ 10 ❺ 20

❻ **a** 3 **b** 7

Challenge

Answers will vary.

Activity

10, 15, 20, 25, 30, 35, 40, 45, 50, 55
11, 16, 21, 26, 31, 36, 41, 46, 51, 56

27:1

❶ 4 ❷ 4 ❸ 2 ❹ 5 ❺ 3

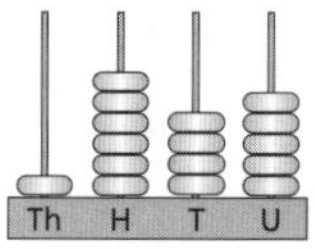

❻ 3 ❼ 10 ❽ 10 ❾ 8321

❿ **a** (abacus: Th H T U) **b** 1644 ⓫ **a** 16, 8, 8 **b** 10, 2, 2

⓬ 3, 3 ⓭ 4 cm, 40 mm

⓮ **a** 5, 5 **b** 5, 5 **c** 9, 9 **d** 9, 9

⓯ **a** 3 spaceships will be coloured. **b** $\frac{5}{8}$ **c** no ⓰ 2

27:2

❶ 8 ❷ 8 ❸ 9 ❹ 9 ❺ 89 ❻ 6 ❼ 6 ❽ 8 ❾ 8

❿ 42c ⓫ 1569 ⓬ 1 2 5 tens 6 ones

⓭ 7, 7 ⓮ **a** 7000 (or 7 thousands) **b** 70 (or 7 tens)

⓯ **a** 24, 8, 8 **b** 24, 6, 6 ⓰ five thousand one hundred and ninety

⓱ 459 ⓲ 2 cm, 20 mm ⓳ 3

Activity

(1) metres (2) centimetres (3) millimetres (4) litres
(5) kilograms (6) minutes (7) centimetre (8) digits

27:3

❶ **a** 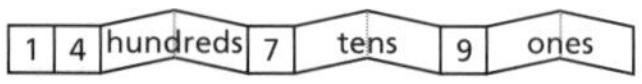 1 4 hundreds 7 tens 9 ones

b  1 4 7 tens 9 ones ❷ 1062 ❸ 1999

❹ 6789 ❺ 819 ❻ **a** 35 mm **b** 16 mm **c** 43 mm

❼ 2002

❽ four thousand and twenty ❾

❿ **a** 38 **b** 45

27:4

❶ 60c ❷ 10 ❸ **a** $1.85 **b** 55c ❹ 97 ❺ 24

❻ 0 ❼ 140 ❽ **a** 91 **b** 90 **c** 92 **d** 92

Challenge

a 6, 20, 20, 10, 9, 10
b 60, 25, 40, 12, 10, 27

Activity

a 6, 6, 6 **b** 7, 7, 7 **c** 7, 7, 7 **d** 9, 9, 9

28:1

❶ 10 ❷ 10 ❸ 2 ❹ 2 ❺ 68 ❻ 2 ❼ 2 ❽ 2
❾ 2 ❿ 35c ⓫ 1324, 3599 ⓬ 1234
⓭ 9 tens and 8 ones (or 98)
⓮ $44 ⓯ **a** 39 **b** 47 ⓰ 3

28:2

❶ 7 ❷ 10 ❸ 4 ❹ 3 ❺ 89 ❻ 79 ❼ 60 ❽ 67
❾ 16 ❿ 43c ⓫ 5243 ⓬ 6 cm, 60 mm ⓭ 59 ⓮ 13
⓯ **a** cone **b** yes **c** yes ⓰ 6 ⓱ **a** 4439 **b** 3676

Activity

(20) net of a cube (21) corner (22) edge (23) face
(24) cube (25) prism (26) pyramid (27) base
(28) cylinder (29) cone (30) sphere

28:3

❶ 9, 9 ❷ 3545 ❸ 4052 ❹ 302
❺ **a** 1 cm 8 mm **b** 2 cm 8 mm
❻ 3 tens and 8 ones (or 38) ❼ 12
❽ cylinder, circles, circle ❾ **a** 77 **b** 74 ❿ 5476

28:4

❶ **a** **B**
❷ **a** 100 cm **b** 100 mm **c** 500 mm **d** 1000 mm
❸ 25 ❹ 10 ❺ 3

Challenge

Answers will vary. One answer could be:
This is called a triangular pyramid. It has 4 corners. 4 flat faces. 6 edges. It can slide. It cannot roll or stack. It has a triangular base.

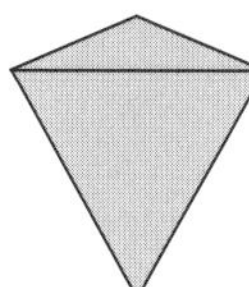

Activity

3, 6, 1, 9, 5, 8, 0, 7, 4, 10
3, 1, 7, 5, 2, 8, 6, 10, 4, 9

29:1

❶ 55 ❷ 69 ❸ 45 ❹ 46 ❺ 31 ❻ 60 ❼ 15 ❽ 7
❾ 12 ❿ 70 ⓫ **a** 8 **b** 6 **c** triangular prism
d B, C, F **e** D, E, H ⓬ tally marks ⓭ 28 + 7 = 35
⓮ 2421 ⓯ yes

29:2

❶ 64 ❷ 81 ❸ 50 ❹ 52 ❺ 93 ❻ 40 ❼ 15
❽ 25 ❾ 18 ❿ 106
⓫

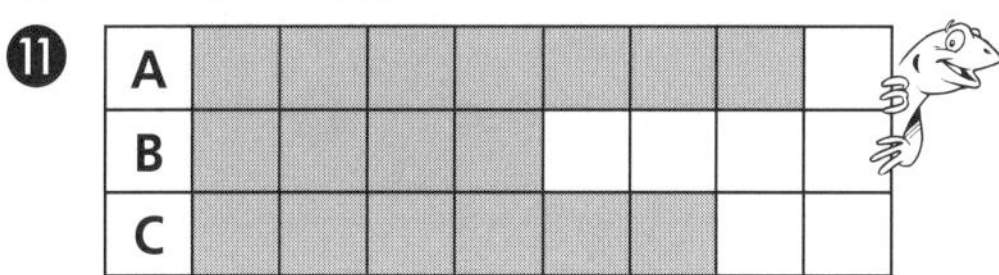

⓬ cube, 6, 12, 8, square ⓭ 46 + 36 = 82 ⓮ **a** 5 **b** 5
⓯ 6 thousands + 4 tens + 3 ones ⓰ 10

Activity

68, 61, 42, 57, 75, 21, 26, 55, 78, 72
69, 62, 43, 58, 76, 22, 27, 56, 79, 73

29:3

❶ 43 ❷ 62 ❸ 59 + 46 =105 ❹ **a** 5 cm **b** 6 **c** 26
❺ rectangular prism, 6, 12, 8, rectangle
❻ rectangular pyramid, rectangular prism

29:4

❶ square pyramid

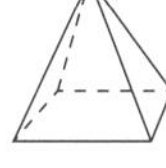

❷ 52 + 39 = 91
❸  ❹ 18

Challenge

Answers will vary.
One answer is:

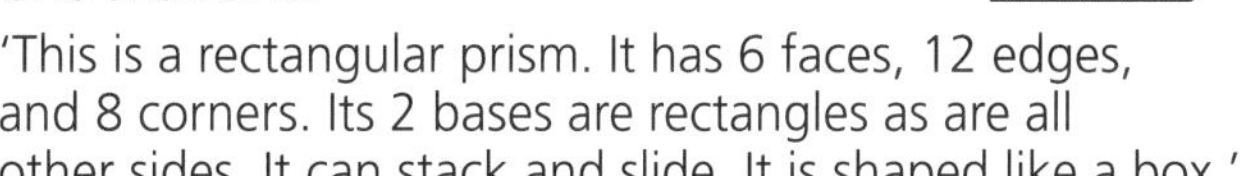

'This is a rectangular prism. It has 6 faces, 12 edges, and 8 corners. Its 2 bases are rectangles as are all other sides. It can stack and slide. It is shaped like a box.'

Activity

70, 63, 44, 59, 77, 23, 28, 57, 80, 74
71, 64, 45, 60, 78, 24, 29, 58, 81, 75

30:1

❶ 5 ❷ 3 ❸ 4 ❹ 1 ❺ 419 ❻ 72 ❼ 6 ❽ 50
❾ 25 ❿ 519 ⓫ **a** 10 **b** 15 ⓬ 29 + 15 = 44 ⓭ no
⓮

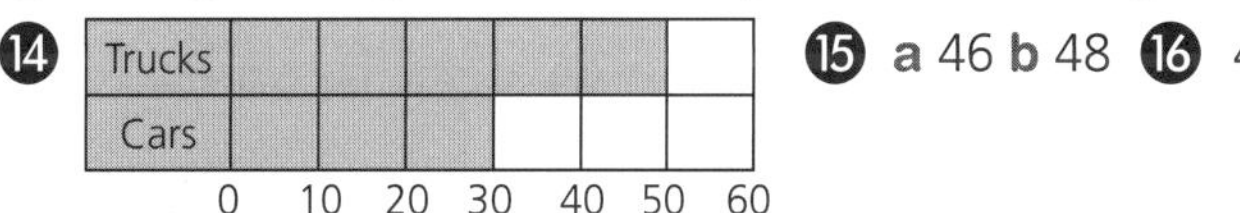

⓯ **a** 46 **b** 48 ⓰ 4

30:2

❶ 8 ❷ 8 ❸ 11 ❹ 6 ❺ 428 ❻ 21 ❼ 32 ❽ 18
❾ 20 ❿ 963 ⓫ **a** unlikely **b** even chance
c very likely **d** certain ⓬ 36 + 38 = 74 ⓭ 500
⓮ **a** 70 **b** 90

Activity

a no **b** even chance **c** less

30:3

1 562 **2** 401 **3** **a** pyramid **b** prism **4** 54

5 **a** Carnation **b** Tulips **c** 8 **d** 8 **e** 30 **6** **a** 61 **b** 52

30:4

1 cube **2** 35 + 18 = 53 **3** yes **4** 95

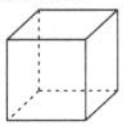

Challenge

Answers will vary and may include: It is an odd number and has 4 digits. It is higher than 3013 and lower than 3034. It rounds to 3030 (to the nearest 10) and 3000 (to the nearest 1000). It can be written as three thousand and twenty-five, 3000 + 20 + 5 or (3 × 1000) + (2 × 10) + (5 × 1). The number before is 3024 and the number after is 3026. It can be represented using 3 thousand blocks, 2 tens blocks and 5 ones blocks. It is in the pattern 3005, 3015, 3025, 3035. 3035 – 10 = 3025, 3005 + 20 = 3025

Activity

a 35 + 37 = 72 **b** 54 + 28 = 82 **c** 60 + 40 = 100

31:1

1 39 **2** 51 **3** 49 **4** 69 **5** 625 **6** 53 **7** 16

8 15 **9** 70 **10** 923 **11** **a** 81, 91, 101, 111

b 60, 65, 70, 75 **c** 12, 15, 18, 21 **d** 460, 470 **12** **A**

13 4 rectangles will be coloured. **14** 7250

15 **a** 62 **b** 90 **16** 10

31:2

1 345 **2** 161 **3** 43 **4** 25 **5** 261 **6** 28 **7** 14

8 12 **9** 60 **10** 913 **11** 45 + 35 = 80

12 **a** grey **b** no **13** 28 **14** **a** 93 **b** 78 **15** 9

Activity

8, 4, 10, 6, 9, 5, 7, 11, 3, 1

9, 5, 14, 7, 10, 6, 8, 12, 4, 2

31:3

1 **a** 90 **b** 104 **2** **B** and **C** **3** Estimates will vary. 13

4 **a** $\frac{3}{5}$ **b** $\frac{2}{5}$ **5** 428 **6** **C**

31:4

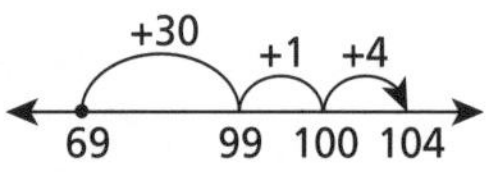

1 yes **2** **a** 174 **b** 177 **3** 156

4 **a** $1\frac{1}{2}$ (number line: 69 +30 → 99 +1 → 100 +4 → 104) **b** 3 **c** 6

5 104 **6** **a** 35 **b** 34

Challenge

Answers will vary. One example is that you may see a crocodile at school tomorrow.

Activity

a total of 7 **b** Answers will vary.

32:1

1 78 **2** 59 **3** 27 **4** 69 **5** 64 **6** 53 **7** 25 **8** 8

9 70 **10** 76 **11** 300 m **12** **a** **B** **b** **C**

c 26 square units **13** 3521 **14** **a** 63 **b** 64

15 30, 35, 40, 45, 50

32:2

1 429 **2** 172 **3** 31 **4** 25 **5** 53 **6** 16 **7** 14

8 12 **9** 60 **10** 39 **11** 25 m

12 **a** 8 forty-seven **b** 47 minutes past 8

13 3, 9cm^2 **14** **a** 61 **b** 81 **15** 3

Activity

a 6148, 6158, 6157

b 8381, 8081, 8061, 8065

c 9027, 9427, 7427, 7422, 7482

32:3

1 33 **2** 51 **3** 77

4 triangular pyramid, 4, 6, 4, triangle

5 4, 8 cm^2 **6** 4:02, 2 **b** 10:36, 36 **7** **a** no **b** no

8 **a** 82 **b** 73

32:4

1 27 **2** 17 **3** 4 **4** **a** 696 **b** 791 **c** 299 **d** 798

Challenge

Answers will vary and could include:

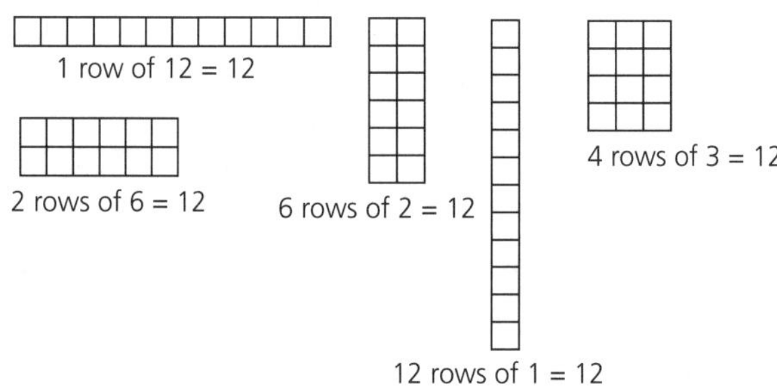

Activity

10, 30, 50, 70, 100, 80, 20, 60, 40, 90

0, 15, 25, 35, 50, 40, 10, 30, 20, 45

2, 6, 10, 14, 20, 16, 4, 12, 8, 18

33:1

❶ 33 ❷ 45 ❸ 3 ❹ 6 ❺ 14 ❻ 42 ❼ 18 ❽ 9
❾ 90 ❿ 28 ⓫ 80 cm ⓬ 3, 6 cm^2 ⓭ yes
⓮ B, D, C, A ⓯ data (information) ⓰ no ⓱ 7

33:2

❶ 79 ❷ 32 ❸ 27 ❹ 69 ❺ 35 ❻ 36 ❼ 15
❽ 30 ❾ 60 ❿ 25 ⓫ 34 cm ⓬ 4, 12 cm^2
⓭ **a** 81, 45 **b** 65, 27 **c** 65, 39
⓮ **a** even chance **b** impossible **c** unlikely **d** certain

Activity

3, 9, 15, 21, 30, 24, 6, 18, 12, 27
0, 12, 20, 28, 40, 32, 8, 24, 16, 36
5, 15, 25, 35, 50, 40, 10, 30, 20, 45

33:3

❶ 38 ❷ 48 ❸ 68 ❹ $97 ❺ $84 ❻ 81, 52
❼ **a** Diane **b** 12 **c** 11 ❽ 7 cm, 70 mm ❾ no
❿ **a** 78, 52 **b** 81, 35

33:4

❶ 90 cm ❷ 55 ❸ **a** 13 km **b** 37 km **c** 24 km ❹ **A**

Challenge

Answers will vary.

Activity

84, 61, 48, 46, 42, 59, 75, 43, 57, 40
82, 59, 33, 64, 50, 57, 83, 61, 65, 48

34:1

❶ 5 ❷ 2 ❸ 7 ❹ 6 ❺ 12 ❻ 24 ❼ 10 ❽ 8
❾ 15 ❿ 18 ⓫ **a** $\frac{4}{10}$ **b** $\frac{7}{10}$ **c** $\frac{9}{10}$
⓬ **a** R1, R3 **b** the Staff room **c** the Canteen
d the Office and Library **e** the Toilets
⓭ **a** 53, 63, 73, 83 **b** 65, 60, 55, 50

34:2

❶ 14 ❷ 15 ❸ 12 ❹ 17 ❺ 15 ❻ 38 ❼ 28
❽ 16 ❾ 40 ❿ 26 ⓫ **a** 14 **b** 8 ⓬ 60 cm
⓭ **a** 0·3 **b** 0·5 **c** 0·8 **d** 1·2 **e** 3·7
⓮ **a** 70, 45 **b** 53, 24

Activity

(8) rhombus (or diamond) (9) trapezium
(10) parallelogram (11) quadrilaterals (12) pentagons
(13) hexagons (16) regular shapes (17) irregular shapes

34:3

❶ 18 ❷ $26 ❸ **a** Yan **b** 6 **c** 4 **d** 20
❹ The 'Total' column is: 22, 8, 19, 28 **b** possum
❺ 0·6 ❻ 0·9

34:4

❶ 6 ❷ **a** 10:00 **b** 12:13 ❸ 6 ❹ 5

Challenge

Answers will vary.
a 6028, 6128, 6118 **b** 9341, 9541, 6541, 6511, 6513
c 5033, 5833, 2833, 2837, 2877

35:1

❶ 16 ❷ 20 ❸ 30 ❹ 30 ❺ 49 ❻ 8 ❼ 19 ❽ 45
❾ 27 ❿ 19 ⓫ **a** 0·4, 0·5, 0·6 **b** $\frac{4}{10}, \frac{5}{10}, \frac{6}{10}$ ⓬ 2·6
⓭ 0·6 ⓮ 34 321 ⓯ 465 780
⓰ Answers could include:
It is a sphere. It has one curved surface, no edges, no corners. It can roll.

35:2

❶ 14 ❷ 40 ❸ 60 ❹ 35 ❺ 23 ❻ 31 ❼ 83
❽ 72 ❾ 7 ❿ 18 ⓫ **F** ⓬ **a** $\frac{3}{10}$ **b** $\frac{6}{10}$ **c** $\frac{9}{10}$
⓭ **a** $1\frac{8}{10}$ **b** $3\frac{9}{10}$ **c** $6\frac{7}{10}$ ⓮ 830 871 ⓯ 7

Activity

(21) abacus (22) table (23) picture graph (24) tally
(25) column graph (26) sector graph (27) digital watch
(28) analog clock (29) calendar (30) balance scales

35:3

❶ **a** 73 741 **b** 90 639 ❷ **a** $\frac{2}{10}$ **b** $\frac{8}{10}$ **c** $\frac{1}{10}$
❸

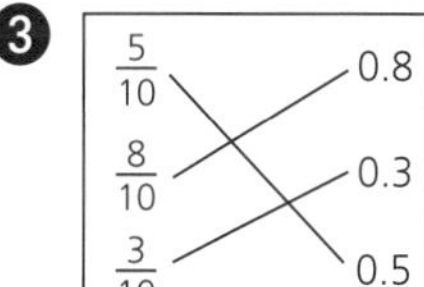

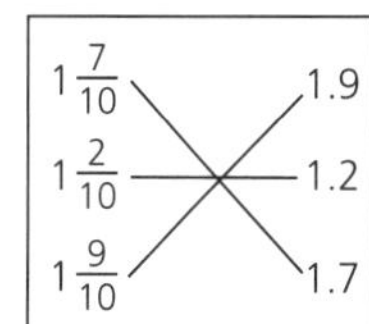

❹ 4.1 ❺

	Thousands	Hundreds	Tens	Units
a	567	6	0	3
b	380	6	8	9
c	293	0	0	4

❻ no

35:4

❶ **a** 2 cm **b** 6 cm

2 6 faces + 8 corners – 12 edges = 2

3 6 560 963 **4** 1 070 630 **5** 7004

Challenge

Answers will vary.

Activity

(20) net of a cube (24) cube (25) prism (26) pyramid (28) cylinder (29) cone (30) sphere

	Faces	Edges	Corners
Number	6	12	8

36:1

1 20 **2** 30 **3** 40 **4** 50 **5** 67 **6** 2 **7** 3 **8** 12 **9** 13 **10** 39 **11** **a** $\frac{2}{10}$ **b** $\frac{4}{10}$ **c** $\frac{8}{10}$ **12** 5 901 657 **13** 1·2 **14** cube and cylinder **15** **a** 41 **b** 33 **c** 63 **d** 72

36:2

1 45 **2** 145 **3** 24 **4** 38 **5** 29 **6** 748 **7** 458 **8** 84 **9** 64 **10** 13 **11** **a** $2\frac{3}{10}$ **b** $4\frac{5}{10}$ **c** $7\frac{2}{10}$ **12** prism **13** 6 **14** no **15** 95 **16** **a** 153 **b** 142

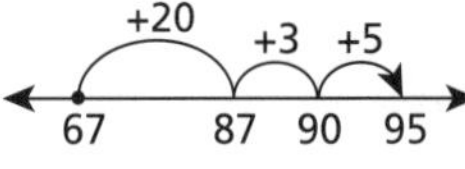

Activity

4, 7, 10, 3, 8, 5, 12, 11, 6, 9
14, 8, 11, 5, 9, 6, 13, 4, 7, 10

36:3

1 2·6 **2** **a** $\frac{5}{10}$ **b** $\frac{9}{10}$ **c** $\frac{2}{10}$ **3** yes

4 **a** 57 **b** 52 **c** 98 **d** 42

5 **a** 85 **b** 38

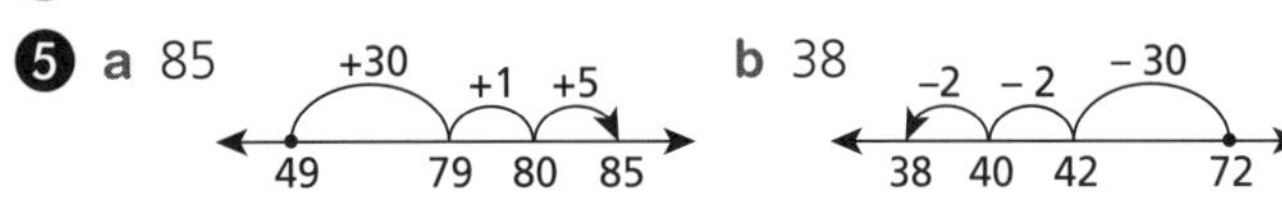

6 **a** circle **b** square **c** circle

36:4

1 6 500 006 **2** **a** 6 **b** 11 **c** 51 **3** **a** 15 **b** 27

4 **a** black rectangle **5** 36

Challenge

Answers will vary. One answer could be:

This is called a triangular prism. It has 6 corners, 5 faces, 9 edges and
2 triangular bases. All other faces are rectangles. It can stack and slide. A Toblerone chocolate bar is the shape of a triangular prism.

Activity

0, 3, 5, 8, 4, 6, 2, 7, 3, 9
7, 3, 10, 2, 6, 8, 4, 9, 5, 1

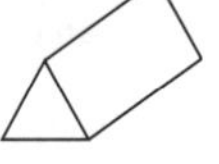

even – even = even
odd – even = odd

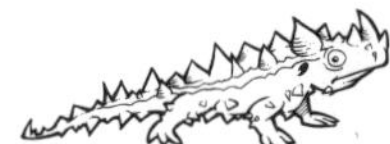

37:1

1 23 **2** 33 **3** 43 **4** 53 **5** 38 **6** 4 **7** 14 **8** 16 **9** 6 **10** 48 **11** **a** **D** **b** **A** **12** The 20c and 10c coins will be coloured. **13** $15 **14** **a** 18th December (or December 18) **b** 13th December (or December 13)

37:2

1 25 **2** 35 **3** 45 **4** 55 **5** 33 **6** 16 **7** 13 **8** 16 **9** 35 **10** 36 **11** $1.20

12 The 50c and 5c coins will be coloured.

13 **a** pansies **b** 15 **c** 5 **14** **a** 603 079 **b** 906 051

Activity

10, 2, 6, 9, 4, 8, 12, 5, 11, 7
8, 11, 3, 6, 12, 5, 10, 13, 7, 9
9, 12, 4, 7, 13, 6, 11, 14, 8, 10

37:3

1 77 **2** 31 **3**

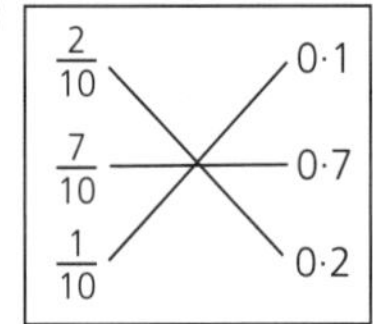

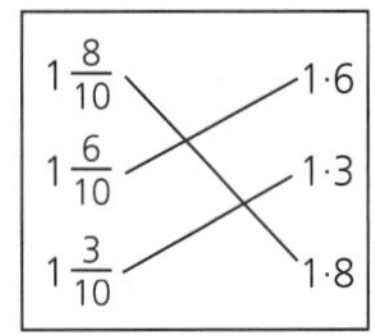

4 The $1, 20c, 10c and 5c coins will be coloured.

5 **a** 6 **b** Emu **c** 4 **6** 40

37:4

1 10 **2** 25 **3** 50 **4** 100 **5** 45 **6** 6 boats will be circled.

7 45, 60, 75, 90, 105 **8** 28

Challenge

(1, 1 and 5), (1, 2 and 4), (1, 3 and 3), (2, 2 and 3)

Activity

Answers will vary.

18:3 out of 11

1. How many legs would there be on 6 dogs? ______

2. Does 4 × 6 = 6 × 4? ______
3. This shape is a ______________________.

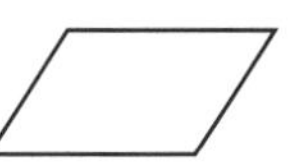

4. Does:
   a 4 + 4 + 4 = 3 × 4? ______
   b 10 × 2 = 2 × 10? ______
   c 78 + 34 = 34 + 78? ______
5. _____ rows of _____ = _____

   _____ × _____ = _____

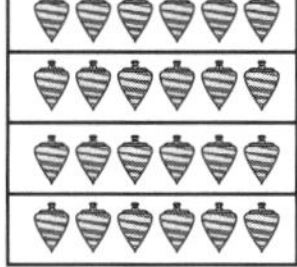

6. Is 3 × 5 half of 3 × 10? ______
7. What numbers could I roll on a standard 6-sided die?

   ____

8. Is it likely that the next school bus you see will have passengers on it? ______

9. What fraction has been coloured?

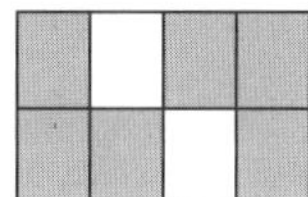

10. 4 × 2 = ____ 2 × 4 = ____ 4 × 4 = ____
11. Draw 4 rows of 5 balls. 4 × 5 = _____

18:4 out of 5

Extension

1. This pyramid is made of 3 layers of 10c coins. What is the total value if it is:
   a 3 layers high? ______
   b 7 layers high? ______
2. 1 4 2 4 3

   Two of these cards are to be drawn from a hat and added. What is the most likely total? ______
3. 5 pentagons have _______ sides.
4. | 1 | 3 | 9 | 10 | 12 | 7 |
   |---|---|---|---|---|---|

   Which three of these numbers add to give:
   a 17? ____________ b 19? ____________
5. 6 △ 8 = 14 △ = ______

Challenge

Draw a 2D shape and describe it.

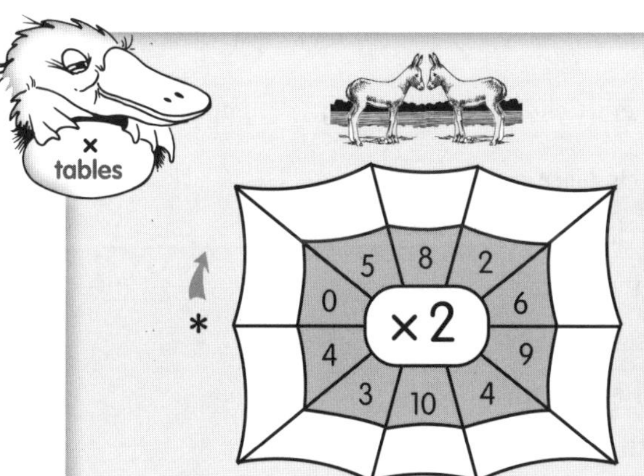

6 groups of 2 is the same as **3 rows of 4**.

	2	4	6	8	10
× 2					

	1	2	3	4	5
× 4					

19:1 out of 16

1. 2 × 4 ____
2. 3 × 4 ____
3. 10 × 4 ____
4. 5 × 4 ____
5. 27 + 12
6. Double 9. ____
7. Half of 18. ____
8. 33 + 22 ____
9. 44 + 11 ____
10. \$56 − \$11
11. Start at 40. Count by 4s.
    40, ____, ____, ____, ____, ____
12. Six teams of 4 children.
    How many altogether? ____
13. 7 △ 3 = 21 △ = ____
14. Follow the rule to continue each pattern.
    - a Add 5. 15, ____, ____, ____
    - b Subtract 2. 20, ____, ____, ____
    - c Add 10. 4, ____, ____, ____
    - d Subtract 10. 60, ____, ____, ____
15. 
    - a A triangle has ____ angles.
    - b A rectangle has ____ angles.
16. Which angle is the smallest? ____

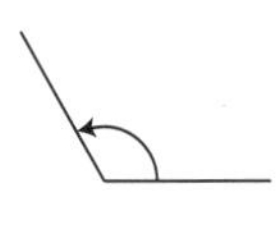
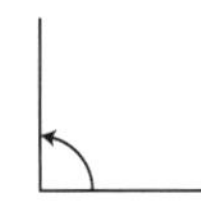

A B C

19:2 out of 15

1. 6 × 4 ____
2. 8 × 4 ____
3. 9 × 4 ____
4. 6 × 4 ____
5. 53 + 41
6. 30 take away 3. ____
7. Add 25 to 34. ____
8. 48 minus 35. ____
9. 45 plus 44. ____
10. \$44 − \$34
11. True or false?

    - a 3 × 9 = 9 × 3 ____
    - b 3 × 9 = 9 + 9 + 9 ____
12. 5 flowers in each bunch.
    How many in 4 bunches? ____
13. a 28, 24, 20, ____, ____, ____, ____
    - b 76, 66, 56, ____, ____, ____, ____
    - c 64, 54, 44, ____, ____, ____, ____
    - d 71, 73, 75, ____, ____, ____, ____
14. In a pentagon, is the number of sides equal to the number of angles? ____
    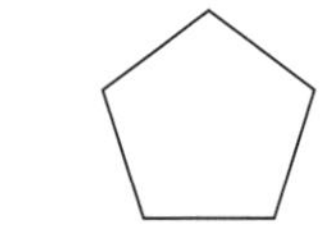
15. Use a ruler to copy this angle.
    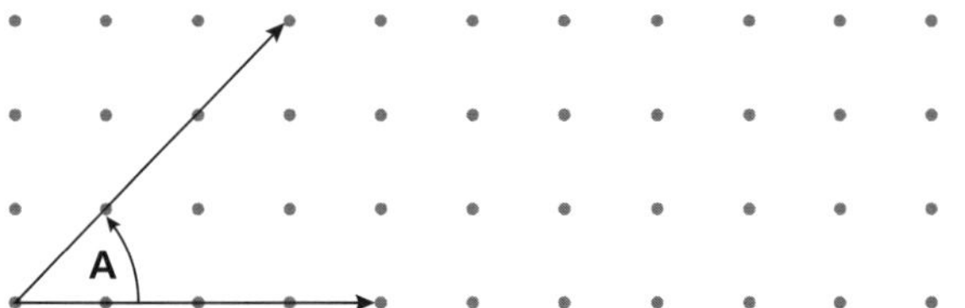

Chance

If something will **always** happen then it is **certain**.

If something will **never** happen then it is **impossible**.

certain | likely | even chance | unlikely | impossible

Choose one of the cards above for each part.

- a It will rain on Saturday. ____
- b I will get wet when I swim. ____
- c I will fall if I climb a tree. ____
- d I will grow a tail if I eat eggs. ____
- e If I toss a coin, it will be a head. ____

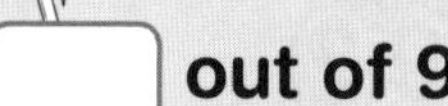

19:3 out of 9

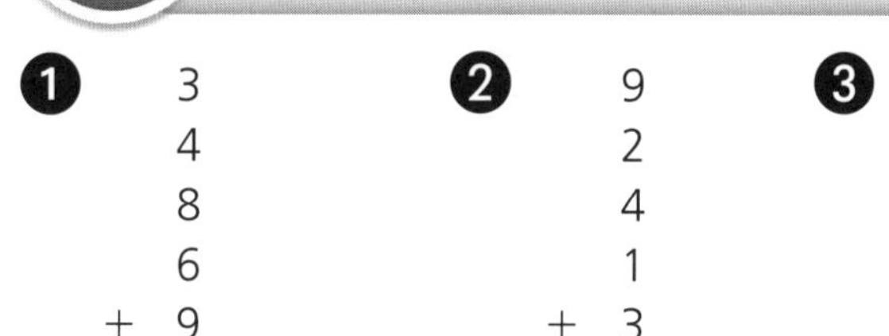

1
$$\begin{array}{r} 3 \\ 4 \\ 8 \\ 6 \\ +\ 9 \\ \hline \end{array}$$

2
$$\begin{array}{r} 9 \\ 2 \\ 4 \\ 1 \\ +\ 3 \\ \hline \end{array}$$

3
$$\begin{array}{r} 8 \\ 6 \\ 2 \\ 1 \\ +\ 4 \\ \hline \end{array}$$

4 Multiply by 2 to complete these patterns.

a 2, 4, 8, ______, ______, ______

b 3, 6, 12, ______, ______, ______

c 4, 8, 16, ______, ______, ______

d 50, 100, ______, ______, ______

5 Which angle is:

a smallest? ______ **b** largest? ______

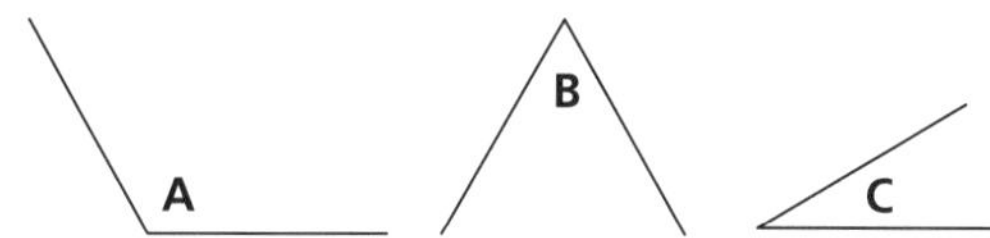

6 Start at 40 and count by 5s.

40, ______, ______, ______, ______,

7 This is called a

r________ angle.

8 Does $5 \times 4 = 4 \times 5$? ______

9 **a** 10 dogs have ______ legs.

b 5 dogs have ______ legs.

c 7 dogs have ______ legs.

d Do 10 dogs have twice as many legs as 5 dogs? ______

19:4 Extension out of 4

1 What shape would you get if you fold the shape along the dotted line?

a ______ **b** ______

2

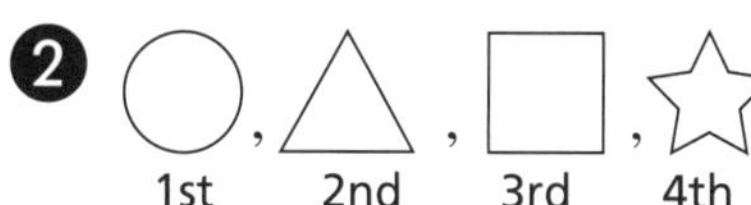

If this pattern were repeated, what would be the 13th shape? ______

3 **a** How many quarters in 5 oranges? ______

b $\frac{1}{4} + \frac{1}{4} + \frac{1}{4} + \frac{1}{4} + \frac{1}{4} + \frac{1}{4}$ ______

c I had 3 oranges cut into quarters. I ate 2 pieces. How much is left? ______

4 How many legs on 5 cats and 3 birds? ______

Challenge

Write chance statements about this spinner.

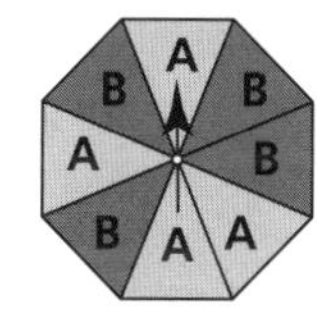

______

______

______

______

______

______

______

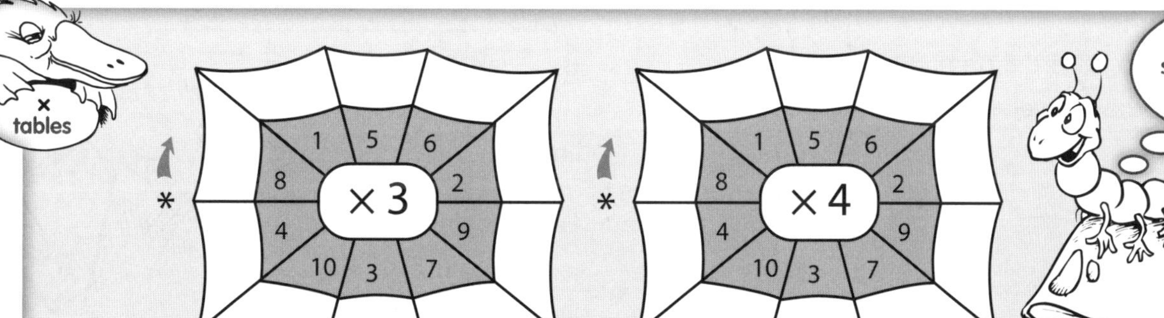

20:1 ☐ out of 19

1. 3×4 ____
2. 4×4 ____
3. 2×3 ____
4. 5×3 ____
5. $\begin{array}{r} 32 \\ +\ 11 \\ \hline \end{array}$
6. 4 less than 12. ____
7. 4 less than 16. ____
8. 3 less than 12. ____
9. 3 less than 9. ____
10. $\begin{array}{r} \$41 \\ -\ \$31 \\ \hline \end{array}$
11. If I toss a coin 10 times, am I very unlikely to get 10 heads? ____
12. Write these angles in order: smallest to largest. ____

A B C

13. If I toss 10 coins, will I always get the same number of heads and tails? ____
14. Does $4 \times 9 = 9 \times 4$? ____
15. I made towers using 3 blocks for each tower. How many blocks would I need for 4 towers? ____
16. 21, 18, 15, ____, ____, ____, ____
17. How many sides on 3 triangles? ____
18. a Draw a trapezium.

    b How many sides? ____

    c How many corners? ____
19. I had $10 and was given $15 more. How much money do I have now? ____

20:2 ☐ out of 17

1. 7×4 ____
2. 9×4 ____
3. 6×3 ____
4. 7×3 ____
5. $\begin{array}{r} 50 \\ +\ 47 \\ \hline \end{array}$
6. 3 less than 21. ____
7. 3 less than 18. ____
8. 4 less than 22. ____
9. 4 less than 33. ____
10. $\begin{array}{r} \$87 \\ -\ \$67 \\ \hline \end{array}$
11. Copy each angle on the dot paper.

A

B

12. Multiply by 10 to complete these patterns.

    a 1, 10, 100 , ____, ____

    b 2, 20, 200, ____, ____

    c 3, 30, 300, ____, ____

    d 4, 40, 400, ____, ____
13. These lines are ____________.
14. When I toss a coin, am I more likely to toss a head than a tail? ____
15. 3, 6, 9, ____, ____, ____, ____, ____
16. Is 4 groups of 3 three more than 3 groups of 3? ____
17. I had $35 and I spent $7. I have ____ left.

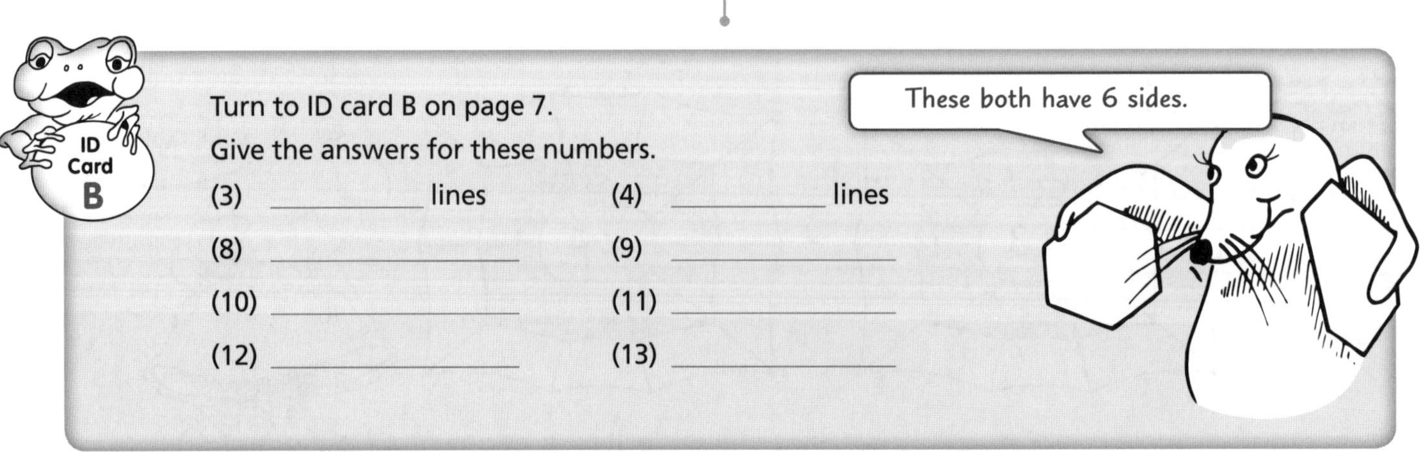

Turn to ID card B on page 7.
Give the answers for these numbers.

(3) ____ lines (4) ____ lines

(8) ____ (9) ____

(10) ____ (11) ____

(12) ____ (13) ____

20:3 ☐ out of 10

1
```
    6
    2
    3
    4
    9
+   1
```

2
```
    5
    4
    3
    2
    1
+   6
```

3
```
    9
    2
    1
    8
    6
+   2
```

4 Is 4 × 10 twice 2 × 10? ______

5 Multiply by 3 to complete these patterns.

a 1, 3, ______, ______

b 2, ______, ______

c 3, ______, ______

d 4, ______, ______

6 Arrange these angles A, B, C and D in order of size, from smallest to largest.

A B C D

______

7 A square corner is called a 'right angle'. Which angle above is a right angle? ______

8 Is there an even chance that I will choose a white ball from the container if I choose without looking? ______

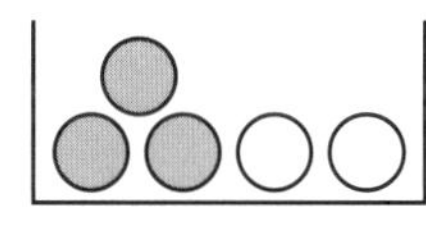

9 I made 7 towers each with 3 blocks. How many blocks did I use? ______

10 **a** How many corners has a circle? ______

b How many corners on 2 triangles? ______

c How many corners on 2 squares? ______

20:4 ☐ out of 5

1 How many pairs of different counting numbers add to give 16? ______

2 How many:

a edges? ______

b faces + corners? ______

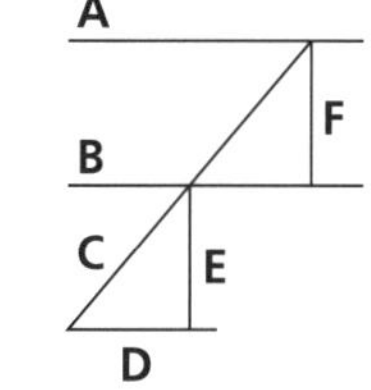

3 **a** The line parallel to **E**. ______

b The lines parallel to **A**. ______

A B C D E F

4 One banana costs 50 cents. How much do four cost? ______

5 This pyramid is made of 3 layers of 5c coins. What is the total value if it is:

a 3 layers high? ______

b 6 layers high? ______

Challenge

Draw a parallelogram and describe it.

Jorge must choose one object without looking.

1 What is the chance that:

a he will pick the black ball? ______

b he will not pick the black ball? ______

c he will choose a cube? ______

d he will choose a ball? ______

2 If I add 10 more black balls to the container, will Jorge have more or less chance of picking the white ball? ______

21:1 out of 14

1. 0 × 3 ____
2. 4 × 3 ____
3. 2 × 4 ____
4. 4 × 4 ____
5. $\begin{array}{r} 66 \\ +\ 11 \\ \hline \end{array}$
6. 3 less than 6. ____
7. 3 less than 3. ____
8. 22 − 11 ____
9. 55 − 11 ____
10. $\begin{array}{r} \$44 \\ -\ \$11 \\ \hline \end{array}$
11. Which lines are parallel? ____

A B C

12. 60 − 5 − 5 − 5 − 5 − 5 − 5 − 5 = ____
13. This is called an ____________.
14. Favourite Shape

Pentagon	⬠ ⬠ ⬠ ⬠
Hexagon	⬡ ⬡ ⬡
Rhombus	◊ ◊ ◊ ◊ ◊ ◊

a What was the most popular shape? ____

b What was the least popular shape? ____

c How many more rhombuses than hexagons? ____

d How many hexagons and pentagons altogether? ____

21:2 out of 19

1. 7 × 3 ____
2. 9 × 3 ____
3. 6 × 4 ____
4. 8 × 4 ____
5. $\begin{array}{r} 39 \\ +\ 50 \\ \hline \end{array}$
6. 4 less than 40. ____
7. 3 less than 30. ____
8. Digits in 4032. ____
9. Half of 24. ____
10. $\begin{array}{r} \$95 \\ -\ \$45 \\ \hline \end{array}$
11. a 30 − 3 − 3 − 3 − 3 = ____

    b 40 − 4 − 4 − 4 − 4 − 4 − 4 = ____
12. Colour two quarters of this shape.

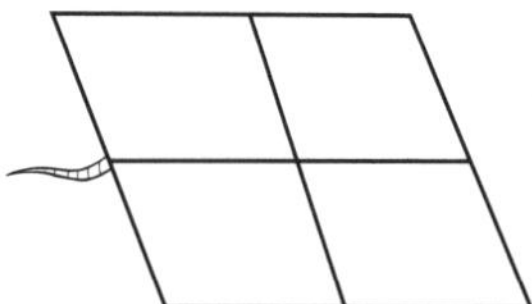

13. Round 44 to the nearest 10. ____
14. Round 546 to the nearest 100. ____
15. Is there an even chance of landing on **A**?

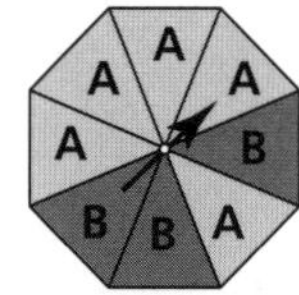

____

16. How many flat surfaces are on a cylinder? ____

17. Write the even numbers between 43 and 48. ____
18. Use the jump strategy to find 64 − 36.

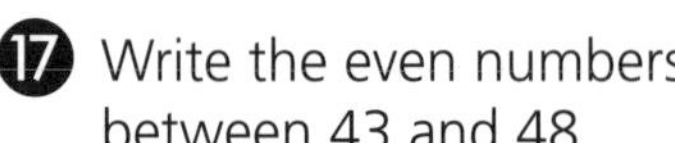

19. Which is larger: 382, 372 or 375? ____

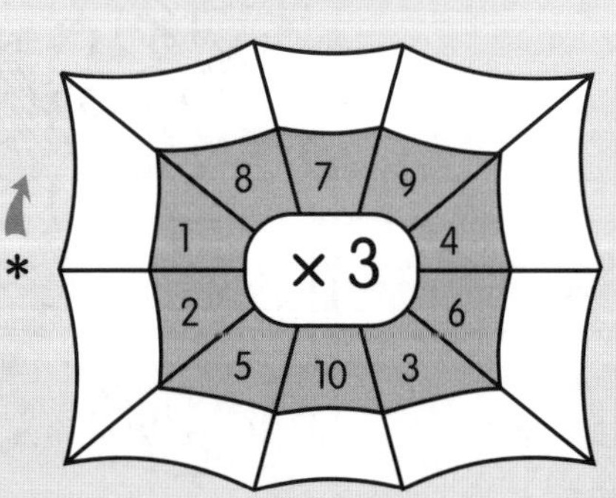

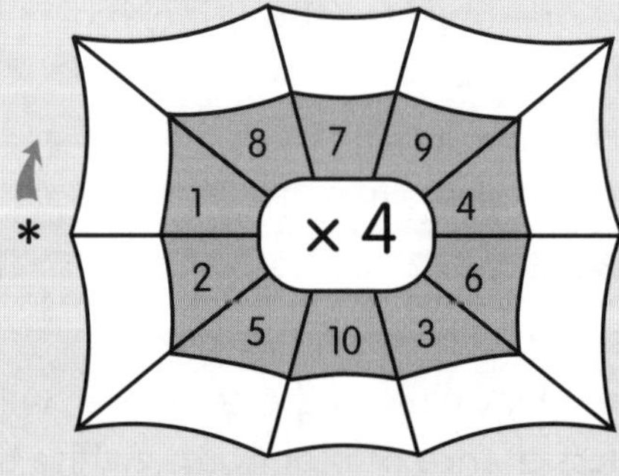

21:3 out of 7

1.

tens	ones
5	6
+3	3

2.

tens	ones
4	1
+5	8

3. Colour 3 fifths of this rectangle.

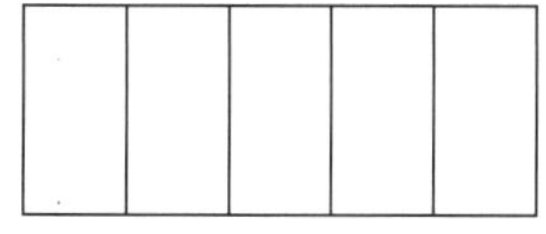

4.     

What part is shaded? ______ out of ______

5.   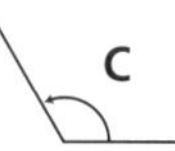

a Which is a right angle? ______

b Which is the largest angle? ______

6.  stands for 4 cars

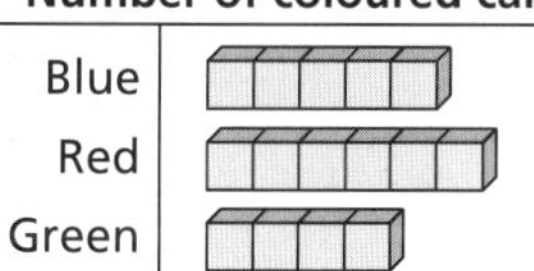

Number of coloured cars	
Blue	
Red	
Green	

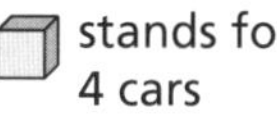

a What coloured car has the greatest number? ______

b How many more red cars than green cars? ______

c How many cars altogether? ______

7. a How many sides on 2 trapeziums? ______

b How many sides on 2 hexagons? ______

c How many sides on 2 pentagons? ______

21:4 out of 5

Extension

1. Colour $\frac{3}{5}$ of these boats blue and $\frac{2}{5}$ of the boats red.

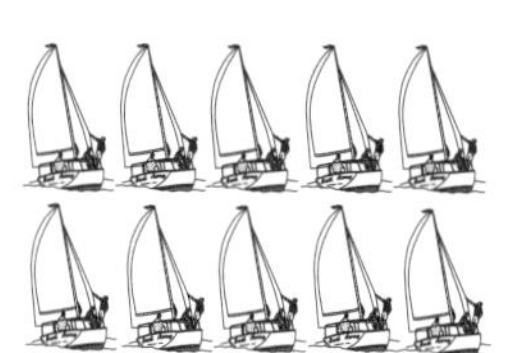

2. The total quantity of drink held in these containers.

 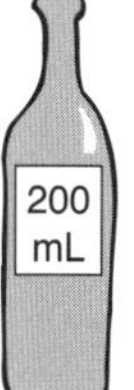 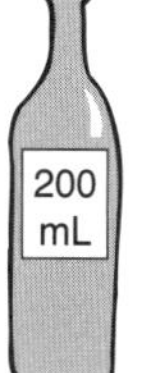 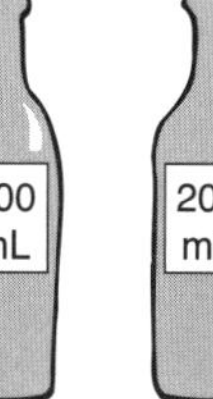

______

3. How many 60c stamps could you buy for $2? ______

4. 38 + 23 ______

38

5. a 40 + 6 + 4 + 8 + 2 + 1 + 7 = ______

b 30 + 17 + 11 + 13 + 19 + 8 = ______

Challenge

Write a pattern with the rule:

a Add 2. ______

b Subtract 3. ______

c Multiply by 2. ______

d Add 8. ______

e Write a pattern using only numbers ending in 0.

______

tables

	0	6	
3			4
	7 –		
7			1
	2	5	

	0	6	
3			4
	8 –		
7			1
	2	5	

22:1 out of 16

❶ 3×3 ____

❷ 6×3 ____

❸ 1×4 ____

❹ 4×5 ____

❺ $\begin{array}{r} 66 \\ +\ 11 \\ \hline \end{array}$

❻ Half of 12. ____

❼ Double 6. ____

❽ Half of 20. ____

❾ Double 11. ____

❿ $\begin{array}{r} \$45 \\ -\ \$31 \\ \hline \end{array}$

⓫ This is called an

__________ .

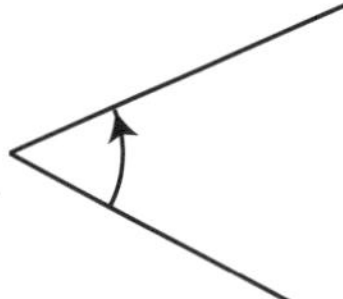

⓬ Am I more likely to pick out a white or grey ball when I take one without looking? ________

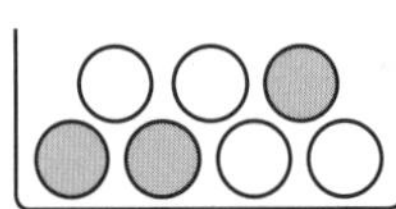

⓭ **a** Colour $\frac{7}{8}$.

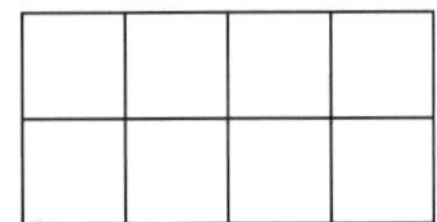

b Colour $\frac{5}{8}$.

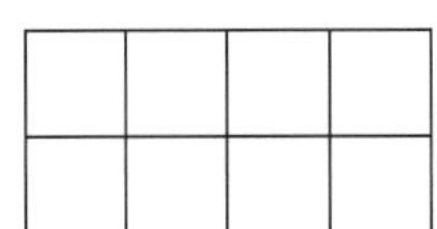

c Which fraction is larger? __________

⓮ The number after 5005 is __________.

⓯ Circle the smallest area.

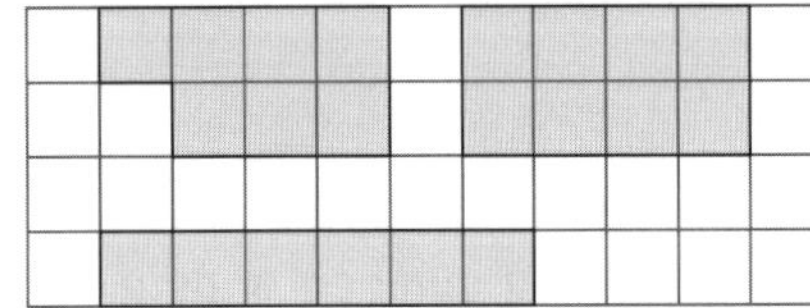

⓰ 4500 = _____ thousands + _____ hundreds

22:2 out of 13

❶ 10×3 ____

❷ 9×3 ____

❸ 8×4 ____

❹ 7×4 ____

❺ $\begin{array}{r} 37 \\ +\ 12 \\ \hline \end{array}$

❻ $54 + 10$ ____

❼ $39 + 10$ ____

❽ $88 - 45$ ____

❾ $79 - 25$ ____

❿ $\begin{array}{r} \$85 \\ -\ \$33 \\ \hline \end{array}$

⓫ **Baskets thrown**

Name	Tally
Marco	𝍸 I
Jill	𝍸 𝍸 II
Kira	IIII
Tim	𝍸

a Who threw the most baskets? ________

b Who threw the least baskets? ________

c How many more did Jill score than Tim? ________

d How many more did the girls score than the boys? ________

e How many baskets were thrown altogether? ________

⓬ Draw a quarter turn anticlockwise.

What number is the hand pointing to now? ________

⓭ **a** A trapezium has __________ angles.

b A hexagon has __________ angles.

tables

11 –	0	9	6	1	3	5	7	2	8	4

12 –	0	9	6	1	3	5	7	2	8	4

12

 ISBN 978 0 6557 0883 4

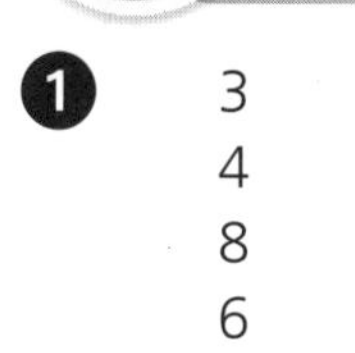

out of 8

❶
```
    3
    4
    8
    6
+   9
```

❷
```
    9
    2
    4
    1
+   3
```

❸
```
    8
    6
    2
    1
+   4
```

❹ 5735 = _____ thousands + _____ hundreds + _____ tens + _____ ones

❺ Write six thousand and thirty-one as a numeral. _____

❻ Use the tally in 22:2 question 11, on the opposite page, to complete this column graph.

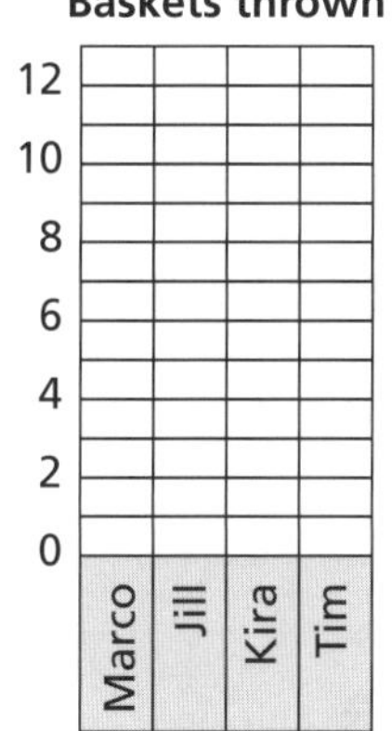

❼ Which angle is larger than a right angle? _____

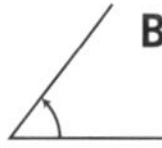

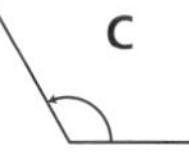

❽ **a** The number shown on the abacus. _____

b How many are in the tens column? _____

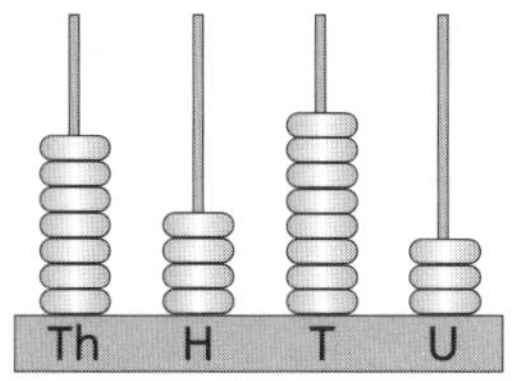

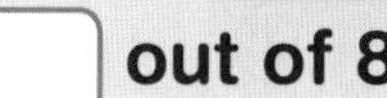

out of 8

❶ The total value of Australia's:

a 2 gold coins. _____

b 4 silver coins. _____

❷ How many hexagons would be in row:

a 4? _____ **b** 10? _____

❸ Colour 1 quarter of this rectangle.

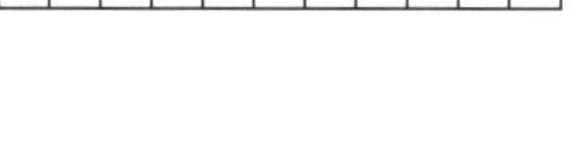

❹ Round off 3239 to the nearest hundred. _____

❺ I have twelve 10 dollar notes.
How much money do I have? _____

❻ Which is larger: $\frac{1}{2}$ or $\frac{2}{3}$? _____

❼ I had 35 apples. I ate some and had 19 left.
How many did I eat? _____

❽ 200 + 5000 + 8 + 20 + 60 = _____

Challenge

Draw a design using trapeziums and/or parallelograms.

out of 16

1. 4×5 ____
2. 2×3 ____
3. $20 - 2$ ____
4. $20 - 5$ ____
5. $57 + 22$ ____
6. Six times two. ____
7. $18 - 9$ ____
8. $5 + $5 + $5 ____
9. $4 + $4 + $4 ____
10. $76 − $22 ____
11. If the clock's hand turns a three-quarter turn clockwise, what does it point to? ____

12. 7294 = ____ thousands + ____ hundreds + ____ tens + ____ ones
13. Which is the largest area? ____

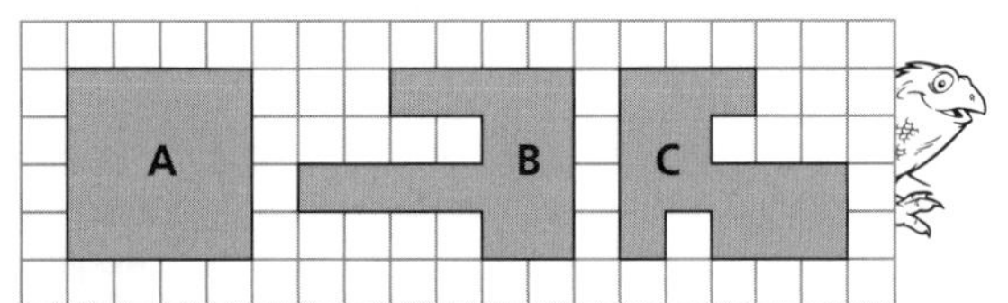

14. Circle the larger fraction.

$\frac{5}{8}$ or $\frac{4}{8}$

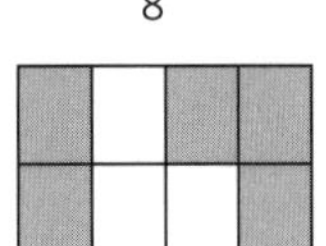
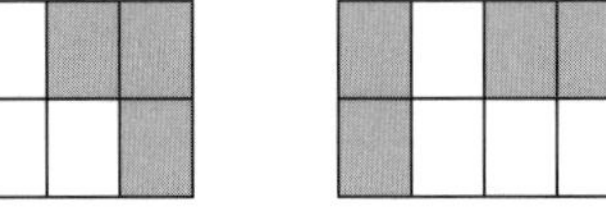

15. How many days in September? ____
16. Days in one week? ____

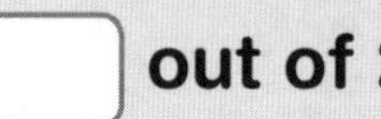

out of 20

1. $40 - 6$ ____
2. $50 - 8$ ____
3. $70 - 3$ ____
4. $90 - 7$ ____
5. $66 + 11$ ____
6. Total of 4 and 3. ____
7. To 24 add 31. ____
8. 23 less than 50. ____
9. Sum of 34 and 4. ____
10. $57 − $15 ____
11. Circle the larger fraction.

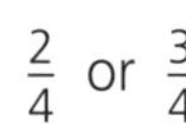

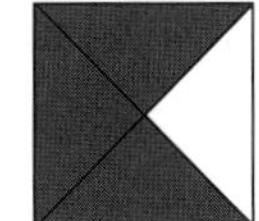

12. 9729 = ____ thousands + ____ hundreds + ____ tens + ____ ones
13. How many days in July and August? ____
14. Name the months in Autumn.

____

____

15. Minutes in 1 hour. ____
16. Days in 1 fortnight. ____
17. Write the short date for 13th April, 2025. ____
18. $3000 + 200 + 70 + 8$ ____
19. Does $4 \times 7 = 7 \times 4$? ____
20. Fill these in for 1479.

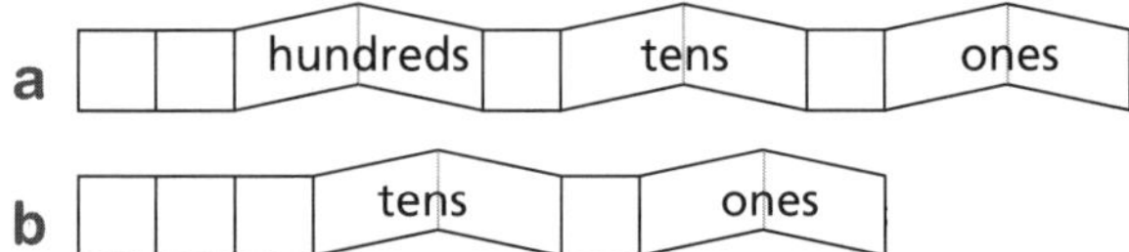

Time

60 minutes	=	1 hour
30 minutes	=	half an hour
7 days	=	1 week
14 days	=	1 fortnight
52 weeks	=	1 year
12 months	=	1 year

Write the months of the year in order.

J ____	F ____	M ____
A ____	M ____	J ____
J ____	A ____	S ____
O ____	N ____	D ____

Underline the months that have 31 days.

23:3 ☐ out of 5

1 a What fraction has been coloured?

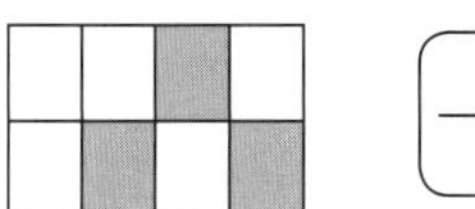

b Colour 4 eighths. Write the fraction.

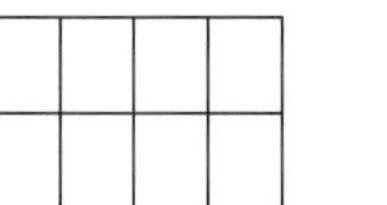

c Tick the larger fraction.

2 a What month is this? ______

October

S	M	T	W	T	F	S
	1	2	3	4	5	6
7	8	9	10	11	12	13
14	15	16	17	18	19	20
21	22	23	24	25	26	27
28	29	30	31			

Note: The calendar for October changes each year.

b How many Tuesdays? ______

c What day is the 12th? ______

d How many days in this month? ______

e What is the date of the first Sunday? ______

f What is the date of the last Saturday? ______

3 How many place-value one blocks do you need to cover this area? ______

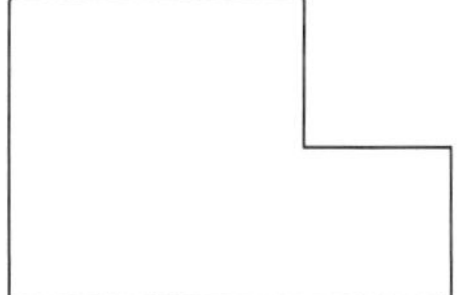

4 8002 = ______ thousands + ______ ones

5 a 3, 6, 9, ______, ______, ______, ______, ______

b 5, 10, 15, ______, ______, ______, ______, ______

23:4 Extension ☐ out of 6

1 Here are 2 desks drawn from above. If 2 sheets of paper can cover one desk, how many sheets are needed to cover each group of desks.

a 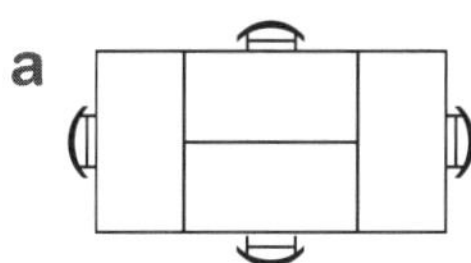 ______

b 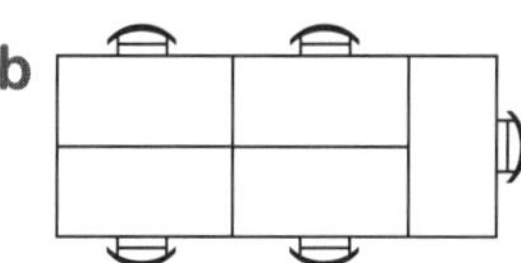 ______

2 This pyramid is made of 20 cent coins.

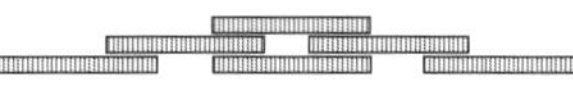

What is the total value if it is:

a 3 layers high? ______

b 5 layers high? ______

3 50 + 2000 + 40 + 1000 + 300 = ______

4 Write a large fraction that is less than 1. $\frac{\square}{\square}$

5 I had 12 oranges. A quarter of them were eaten. How many are left? ______

6 a Sides on 4 pentagons and 3 hexagons. ______

b Sides on 3 kites and 5 parallelograms. ______

Challenge

Draw a path on the grid from the counter to the triangle. Describe the path.

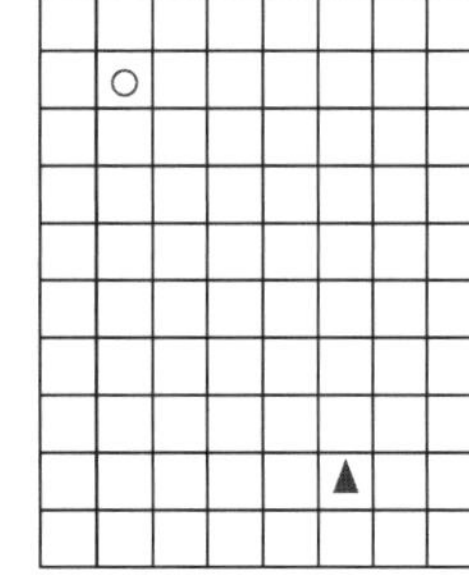

______

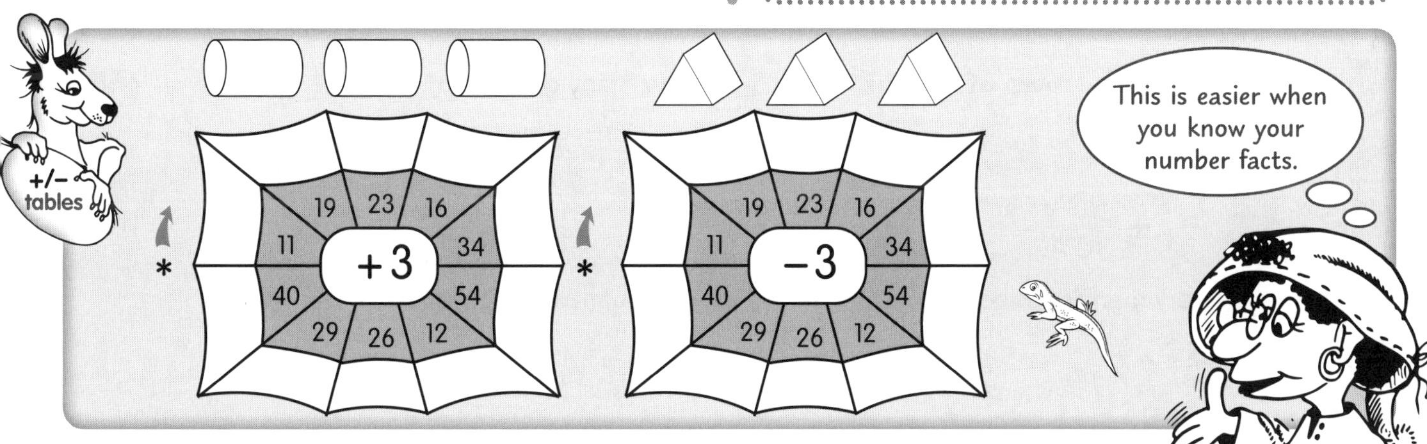

24:1

out of 13

1. 3 × 3 ______
2. 0 × 3 ______
3. 5 × 4 ______
4. 10 × 4 ______
5. Two times five. ______
6. 5c + 5c + 5c ______
7. $2 + $2 + $2 ______
8. Double 9. ______
9. **a** What fraction is shaded? $\frac{\square}{\square}$

   b Write a fraction that is smaller than $\frac{3}{4}$.

10. 

   a Who sits in front of **L**? ______

   b Who sits behind **S**? ______

   c Who sits to the right of **G**? ______

   d Who sits to the left of **P**? ______

   e Who sits in the back right corner? ______

11. 4000 + 900 + 30 + 1 = ______
12.

	3	7	4	5	2	10	6	9
× 2								

13. I had 24 cupcakes. 10 were eaten. How many are left? ______

24:2

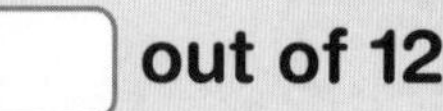

out of 12

1. 3 × 4 ______
2. 0 × 4 ______
3. 5 × 2 ______
4. 5 × 4 ______
5. Two times six. ______
6. 20c − 5c ______
7. $10 − $8 ______
8. Double 4. ______
9. The month 3 months before March. ______
10. **a** How many Mondays? ______

   b The last day is a ______.

   c What day is it 3 days after the 4th of June? ______

June

S	M	T	W	T	F	S
	1	2	3	4	5	6
7	8	9	10	11	12	13
14	15	16	17	18	19	20
21	22	23	24	25	26	27
28	29	30				

Note: The calendar for June changes each year.

11. 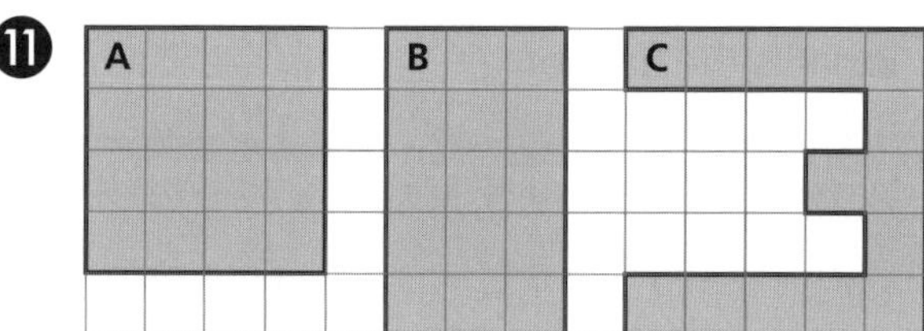

   a Which shape has 15 squares? ______

   b Order these areas from largest to smallest. ______

12. How many people can each be given 4 lollies? ______

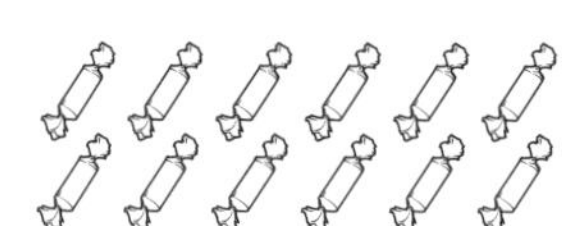

12 ÷ 4 = ______

a How many **rows of 10** coins? ______

b How many **groups of 6** cones? ______

c How many **groups of 4** fish? ______

d How many **groups of 3** stars? ______

 • *AUSTRALIAN SIGNPOST MATHS 3 MENTALS* • ISBN 978 0 6557 0883 4

24:3 ☐ out of 6

1 a Shade $\frac{3}{4}$.

b Shade $\frac{1}{2}$.

c Which is larger $\frac{3}{4}$ or $\frac{1}{2}$? ☐

2 Follow the directions and colour the path of the counter.

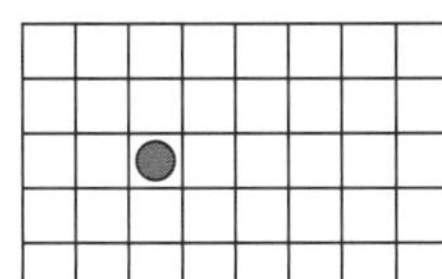

Move 2 up, then 2 left, then 4 down, then 5 right, then 3 up.

3 How many place-value ones blocks are needed to cover this area?

Area = ______ blocks.

4 a Months in one year. ______

b The month before March is ______.

c The month after December is ______.

d November is the ______ month of the year.

e Write the short date for 5th May, 2026. ______

5

	3	7	4	5	2	10	6	9
× 4								

6 a 78, 68, 58, ____, ____, ____, ____

b 20, 22, 24, ____, ____, ____, ____

24:4 ☐ out of 8

Extension

1

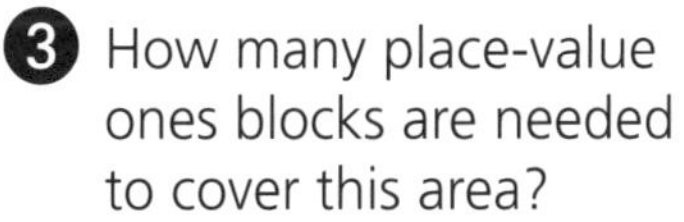

Colour one quarter of the bells red and one half of them blue. What fraction have you coloured? ______

2 What fraction of $1 is:

a 20 cents? ______ b 5 cents? ______

3 70 + 5 + 9000 + 4 + 10 = ______

4 Hours in 3 days. ______

5 Months in 5 years. ______

6 Today is Monday, What day will it be:

a in 6 days time? ______

b 2 weeks from tomorrow? ______

7 a Days in 5 weeks. ______

b Days in 10 weeks. ______

8 I was given $1 on Monday. Each day I was given double the amount I was given the day before.

a How much was I given on Friday? ______

b How much money would I have been given altogether once I got to the 7th day? ______

Challenge

Starting at the circle, colour squares to make a path that ends at the triangle.

Write the directions for the path.

See 24:3, question 2.

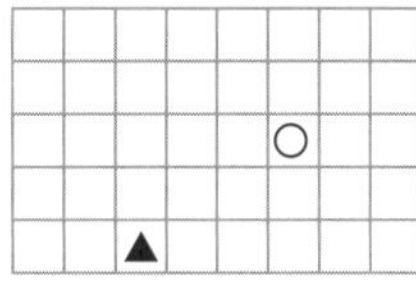

25:1 ☐ out of 15

1. 8 + 12 _____
2. 15 + 5 _____
3. 17 + 3 _____
4. 11 + 9 _____
5. 9 + 9 + 9 + 9 _____
6. 3 + 3 + 3 + 3 _____
7. 20 − 5 _____
8. 20 − 7 _____
9. How many days in:
   a April? _____ b November? _____
   c May? _____ d October? _____
10. How many groups of:

   a 3 hooks? _____
   24 ÷ 3 = _____
   b 4 hooks? _____
   24 ÷ 4 = _____
11. 

   a 2 × 4 = _____
   b 8 ÷ 2 = _____
   c 2 × _____ = 8
12. Do we measure mass with centimetres, grams or litres? _____
13. Use the short form to write thirty grams. _____
14. 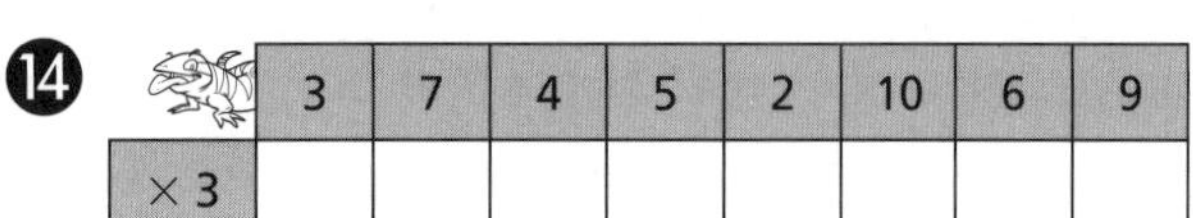

	3	7	4	5	2	10	6	9
× 3								

15. a 65, 60, 55, _____, _____, _____, _____, _____
    b 40, 42, 44, _____, _____, _____, _____, _____
    c 86, 76, 66, _____, _____, _____, _____, _____

25:2 ☐ out of 11

1. 23 − 10 _____
2. 46 + 10 _____
3. 15 + 15 _____
4. 19 + 19 _____
5. Double 29. _____
6. Halve 68. _____
7. Halve 46. _____
8. Double 35. _____
9. How many groups of 5 in 30? _____
   30 ÷ 5 = _____
10. I have 28 toys. To how many children can I give 4 toys?

   _____
   28 ÷ 4 =
11. 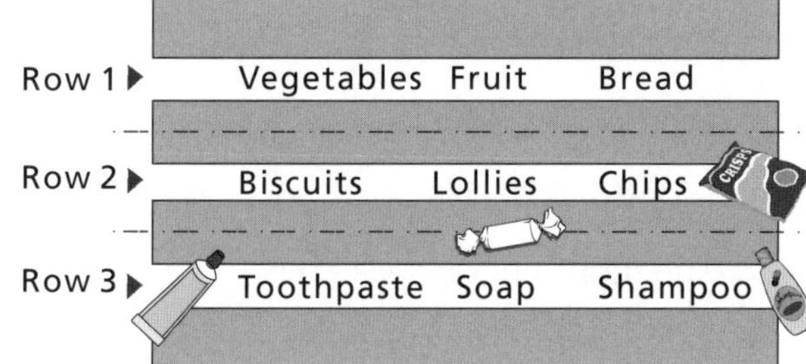

   a Which row has chips in it? _____
   b What is found between bread and vegetables? _____
   c What is found next to the toothpaste? _____
   d Which row has bread in it? _____
   e What is between the biscuits and chips? _____

Concept

Complete the cards.

a 2 rows of 3 = ____ How many 3s in 6? = ____
2 × 3 = ____ so 6 ÷ 3 = ____

b 3 rows of 5 = ____ How many 5s in 15? = ____
3 × 5 = ____ so 15 ÷ 3 = ____

c 3 groups of 3 = ____ How many 3s in 9? = ____
3 × 3 = ____ so 9 ÷ 3 = ____

25:3 ☐ out of 9

1. ______ × 2 = 16 so 16 ÷ 2 = ______

2. How many ears would 10 children have? ______

10 × 2 = ______ 20 ÷ 2 = ______

3. 4 × 10 = ______
40 ÷ 10 = ______
4 × ______ = 40

4. Would a pencil case be measured using grams or kilograms? ______

5. 2 × 3 = ______
6 ÷ 2 = ______
______ × 3 = 6

6. Write the short form for sixty-seven millilitres. ______

7. What is the mass shown on these scales? ______

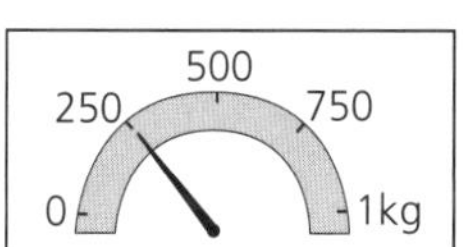

8. What is the volume of water in this container? ______

1000
500
ml

9. Use the jump strategy to find:

47 + 38 = ______

Extension 25:4 ☐ out of 5

1. 5 sheets of paper can make a book.
How many books can I make if I have:
a 20 sheets? ______ **b** 17 sheets? ______

2. How many 500 mL containers would I need to fill a 3 L container? ______

3. I doubled a number and then added 5.
My answer was 23.
What number did I start with? ______

4. A bottle balanced 12 bolts.
A book balanced 4 bolts.
a Does the bottle have the same mass as 12 bolts? ______
b How many books would balance the bottle? ______

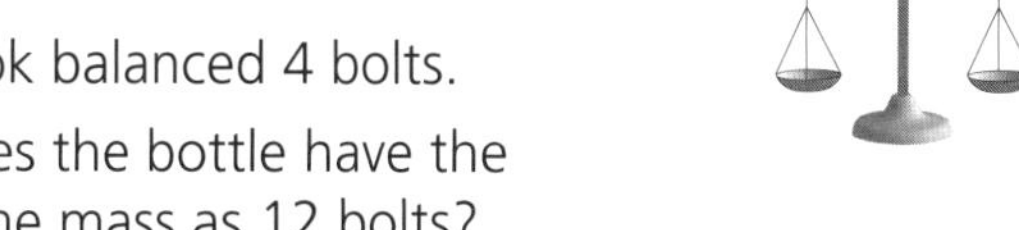

5. 250g + 250g + 250g + 250g = ______

Challenge

Draw and label items you would find in your kitchen pantry that weigh less than 1 kg.

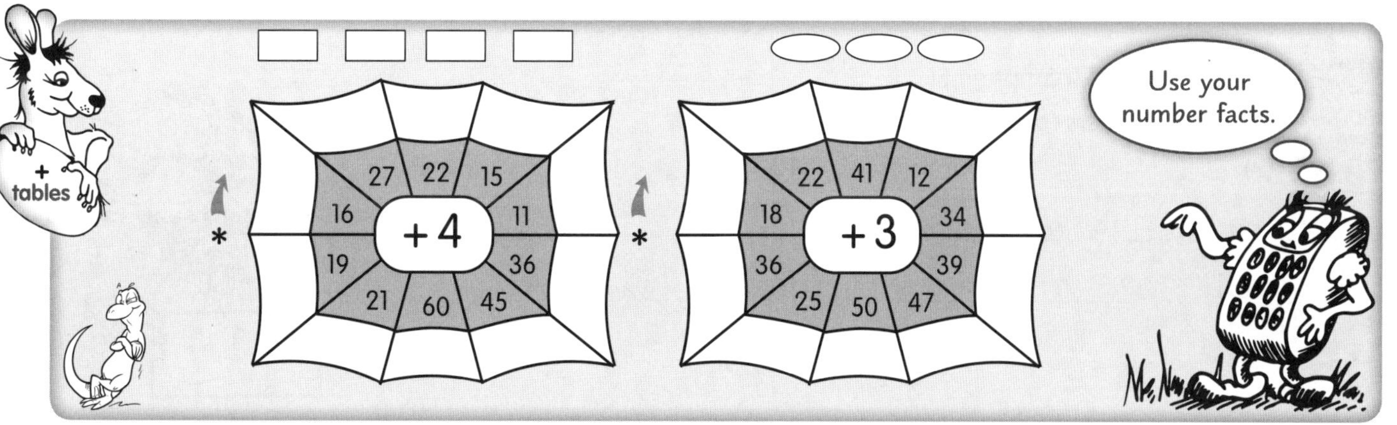

26:1 ______ out of 16

1. ______ × 2 = 12
2. 12 ÷ 2 = ______
3. ______ × 2 = 18
4. 18 ÷ 2 = ______
5. ______ × 10 = 50
6. 50 ÷ 10 = ______
7. ______ × 5 = 15
8. 15 ÷ 5 = ______
9. Circle groups of 5 balls.

   What is 4 groups of 5? ______ 4 × 5 = ______

   How many 5s in 20? ______ 20 ÷ 5 = ______

   Share 20 among 4. ______ 20 ÷ 4 = ______
10. Would a packet of chips weigh 100 grams or 100 kilograms? ______
11. Use the marks on the scale to write the mass to the closest mark shown. ______

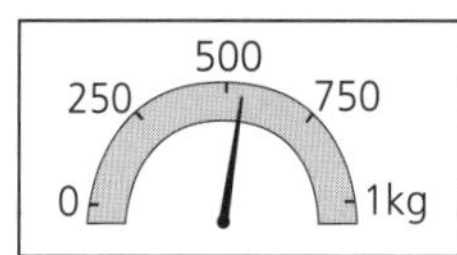

12. Use ______ × 2 = 12 to find 12 ÷ 2 = ______
13. Six people shared 18 pencils. How many did each person get? ______
14. Write twenty-seven millilitres using short form. ______
15. How many days in:

   a one week? ______ b September? ______

   c July? ______ d November? ______
16. 4000 + 300 + 90 + 6 ______

26:2 ______ out of 16

1. ______ × 5 = 20
2. 20 ÷ 5 = ______
3. ______ × 4 = 16
4. 16 ÷ 4 = ______
5. ______ × 10 = 90
6. 90 ÷ 10 = ______
7. ______ × 3 = 21
8. 21 ÷ 3 = ______
9. 

   Circle groups of 4 hamburgers.

   What is 6 groups of 4? ______ 6 × 4 = ______

   How many 4s in 24? ______ 24 ÷ 4 = ______

   Share 24 among 6. ______ 24 ÷ 6 = ______
10. a Half of 18 is ______. b Double 9 is ______.

   c 9 × 2 ÷ 2 = ______
11. One litre is 1000 grams.

   Half a litre is ______ grams.

   A quarter of a litre is ______ grams.
12. What is the volume of water in this container? ______

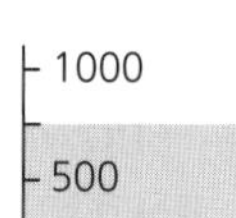

13. Use the jump strategy to find:

   61 − 28 = ______

14. mL stands for ______.
15. 4000 + 200 + 90 + 1 = ______
16. 8 stars shared between 4. How many each? ______

Turn to ID card A on page 6.
Give the answers for these numbers.

(9) ______ numbers (10) ______ numbers

(11) ______ numbers (12) ______

(18) ______ (19) ______

(20) ______ (21) ______

 ISBN 978 0 6557 0883 4

26:3 ___ out of 6

1. Write thirty-nine millilitres using short form. ______
2. How many 500 mL containers would I need to fill a 1 L container? ______
3. Would a student in Year 3 weigh about 28 grams or 28 kilograms? ______
4. **a** 5 × 4 = ____ **b** 6 × 3 = ____

   20 ÷ 4 = ____ 18 ÷ 3 = ____

   5 × 4 ÷ 4 = ____ 6 × 3 ÷ 3 = ____
5. 

   Diane

   Alan

   Pa

   Nanna

   Lyn

   Whose photo is:

   a in the middle? ______

   b at the bottom left? ______

   c at the top right? ______

   d above Nanna's photo? ______

   e below Alan's photo? ______
6. **a** ____ × 5 = 30 so 30 ÷ 5 = ____

   b ____ × 3 = 27 so 27 ÷ 3 = ____

   c ____ × 4 = 32 so 32 ÷ 4 = ____

   d ____ × 10 = 60 so 60 ÷ 10 = ____

26:4 ___ out of 6

Extension

1. If 12 × 9 = 108, find:

   a 108 ÷ 9 ______ **b** 108 ÷ 12 ______
2. If 32 + 19 = 51, then:

   a 19 + 32 = ______ **b** 51 − 19 = ______
3. How many shoes are in 57 pairs? ______

   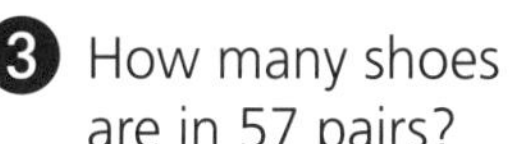

4. How many 500 mL containers would I need to fill a 5 L container? ______
5. Lachlan shared 60 grapes with 2 of his friends. How many did each child get? ______
6. Ten pens fill one box.

   How many boxes can be filled with:

   a 30 pens? ______ **b** 70 pens? ______

Challenge

Draw and label containers that hold less than 1 litre.

+ tables

+5	5	10	15	20	25	30	35	40	45	50
+5	6	11	16	21	26	31	36	41	46	51

Can you see a pattern?

27:1 ☐ out of 16

1. ____ × 3 = 12
2. 12 ÷ 3 ____
3. ____ × 5 = 10
4. 10 ÷ 2 ____
5. ____ × 10 = 30
6. 30 ÷ 10 ____
7. ____ × 2 = 20
8. 20 ÷ 2 ____
9. 8000 + 300 + 20 + 1 = ____
10. **a** Show the number 1645 on this abacus.

    b One less than 1645 is ____________.

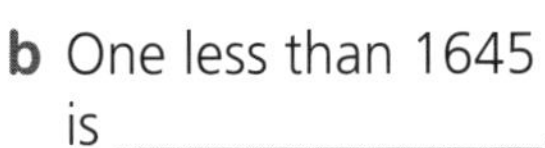

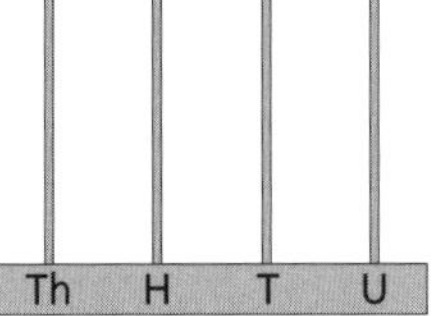

11. **a** 8 × 2 ____

    16 ÷ 2 ____

    8 × 2 ÷ 2 ____

    b 2 × 5 ____

    10 ÷ 5 ____

    2 × 5 ÷ 5 ____

12. Use ____ × 3 = 9 to find 9 ÷ 3 = ____
13. How long is this line: in cm? ______ in mm? ______
14. **a** ____ × 5 = 25 so 25 ÷ 5 = ____

    b ____ × 3 = 15 so 15 ÷ 3 = ____

    c ____ × 4 = 36 so 36 ÷ 4 = ____

    d ____ × 10 = 90 so 90 ÷ 10 = ____

15. **a** Colour three eighths.

    b What fraction is not coloured? $\frac{\square}{\square}$

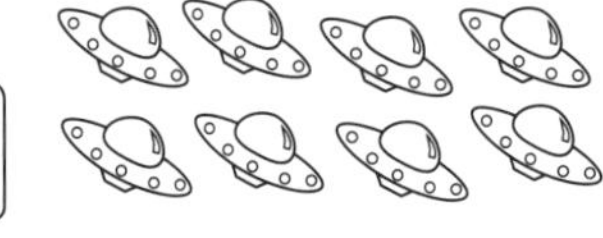

    c Is $\frac{3}{8}$ larger than $\frac{5}{8}$? ____

16. 10 balls shared between 5. How many each? ______

27:2 ☐ out of 19

1. ____ × 5 = 40
2. 40 ÷ 5 ____
3. ____ × 3 = 27
4. 27 ÷ 3 ____
5. $\begin{array}{r} 6\ 3 \\ +\ 2\ 6 \\ \hline \end{array}$
6. ____ × 10 = 60
7. 60 ÷ 10 ____
8. ____ × 2 = 16
9. 16 ÷ 2 ____
10. $\begin{array}{r} 7\ 3\text{c} \\ -\ 3\ 1\text{c} \\ \hline \end{array}$
11. 1000 + 500 + 60 + 9 = ____
12. 1256 = ☐ ☐ ☐ tens ☐ ones

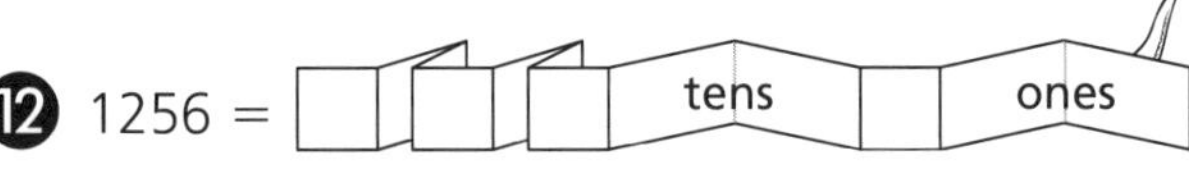

13. Use ____ × 4 = 28 to find 28 ÷ 4 = ____
14. What is the value of the 7 in:

    a 7320? ____ **b** 2072? ____

15. **a** 8 × 3 = ____

    24 ÷ 3 = ____

    8 × 3 ÷ 3 = ____

    b 6 × 4 = ____

    24 ÷ 4 = ____

    6 × 4 ÷ 4 = ____

16. Write 5190 in words.

    ____

    ____

17. I have $4594 in the bank. How many $10 notes could I take out? ______
18. How long is this line: in cm? ______ in mm? ______
19. 15 cupcakes shared between 5 children. How many each? ______

Turn to ID card A on page 6.
Give the answers for these numbers.

(1) ____________ (2) ____________

(3) ____________ (4) ____________

(5) ____________ (6) ____________

(7) ____________ (12) ____________

27:3 ☐ out of 10

1. Fill these in for 1479.

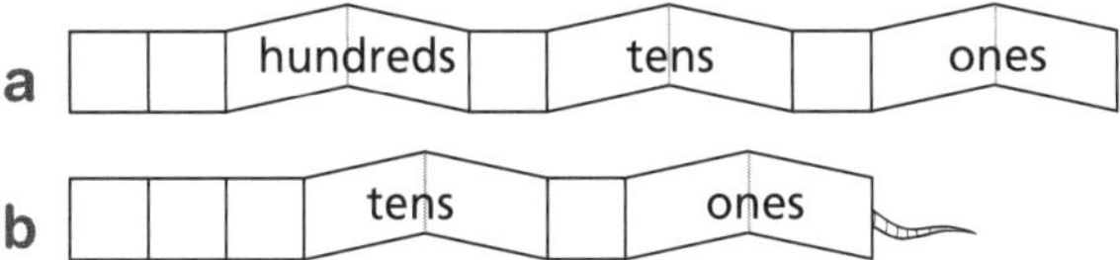

2. Write the numeral for one thousand and sixty-two. ______
3. The year before 2000. ______
4. 6000 + 700 + 80 + 9 = ______
5. I have $8192 in the bank. How many $10 notes could I take out? ______
6. Measure the lines in millimetres.
   a ______
   b ______
   c ______
7. 3 more than 1999. ______
8. Write 4020 in words. ______

______

9. Show the time 3:34 on the clock face.

10.

a	b
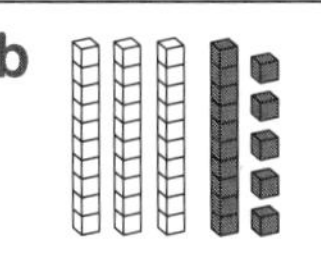 25 + 13 = ______	30 + 15 = ______

27:4 ☐ out of 8

Extension

1. Four identical books cost $2.40 altogether. How much for one book? ______
2. How many hundreds in 1000? ______
3. **A** **B**

   a The total value of the coins in **A** and **B**. ______
   b The difference in value of the coins in **A** and **B**. ______
4. One zoo has 82 monkeys. Another zoo has 15 monkeys. How many monkeys in both zoos? ______
5. Months in 2 years. ______
6. 30 − 6 − 6 − 6 − 6 − 6 ______
7. 20 + 30 + 40 + 50 ______
8. In a non-leap year, how many days are there in:
   a Spring? ______ b Summer? ______
   c Autumn? ______ d Winter? ______

Challenge

Complete each column as quickly as you can. Once finished, check your answers. Record your time.

a		b	
3 × 2	______	6 × 10	______
5 × 4	______	5 × 5	______
2 × 10	______	8 × 5	______
2 × 5	______	4 × 3	______
3 × 3	______	1 × 10	______
5 × 2	______	9 × 3	______

Time = ______ seconds

☐ groups of 6 = 24
☐ × 6 = 24
The answer is 4.

Multiplication linked with division

a ☐ groups of 5 = 30
☐ × 5 = 30
How many groups of 5 in 30? ______.

b 4 shares of ☐ = 28
4 × ☐ = 28
28 shared among 4 boys = ______.

c ☐ rows of 6 = 42
☐ × 6 = 42
How many rows of 6 in 42? ______.

d 3 shares of ☐ = 27
3 × ☐ = 27
27 shared among 3 girls = ______.

28:1

out of 16

1. ____ × 5 = 50
2. 50 ÷ 5 ____
3. ____ × 3 = 6
4. 6 ÷ 3 ____
5. $\begin{array}{r} 45 \\ +\ 23 \\ \hline \end{array}$
6. ____ × 10 = 20
7. 20 ÷ 10 ____
8. ____ × 2 = 4
9. 4 ÷ 2 ____
10. $\begin{array}{r} 5\ 6c \\ -\ 2\ 1c \\ \hline \end{array}$
11. 

    The number modelled above. ____

    The number before 3600. ____
12. 1000 + 200 + 30 + 4 = ____
13. 6 tens and 3 ones
    + 3 tens and 5 ones
14. I paid \$31 on fish and \$13 on bread.

    How much did I spend? ____
15. 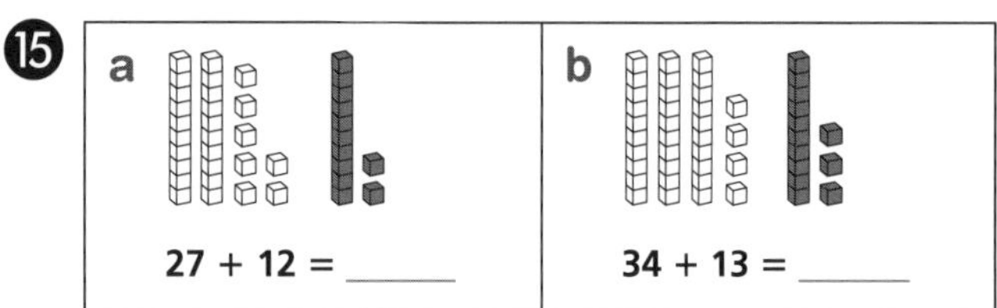

    a 27 + 12 = ____

    b 34 + 13 = ____
16. 12 stickers shared by 4. How many each? ____

28:2

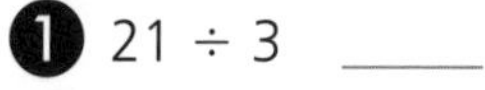

out of 17

1. 21 ÷ 3 ____
2. 50 ÷ 5 ____
3. 40 ÷ 10 ____
4. 9 ÷ 3 ____
5. $\begin{array}{r} 63 \\ +\ 26 \\ \hline \end{array}$
6. 56 + 23 ____
7. 83 − 23 ____
8. 48 + 19 ____
9. 45 − 29 ____
10. $\begin{array}{r} 6\ 7c \\ -\ 2\ 4c \\ \hline \end{array}$
11. 5000 + 200 + 40 + 3 = ____
12. ______________________

    This line is ____ cm long.

    This line is ____ mm long.
13. I saw 34 galahs and 25 cockatoos.

    How many birds did I see? ____
14. 56 forks and 43 knives.

    How many more forks? ____
15. A B C

    a Name solid **B** above. ____

    b Do solids like **A** stack easily? ____

    c Will shape **C** roll? ____
16. A cube has ____ faces.

17. a 100 less than 4539. ____

    b 10 less than 3686. ____

Turn to ID card B on page 7.
Give the answers for these numbers.

(20) ____ (21) ____ (22) ____

(23) ____ (24) ____ (25) ____

(26) ____ (27) ____ (28) ____

(29) ____ (30) ____

28:3 ☐ out of 10

1. Use ____ × 5 = 45 to find 45 ÷ 5 = ____
2. 3000 + 500 + 40 + 5 = ____
3. Write the numeral for the number shown. ____

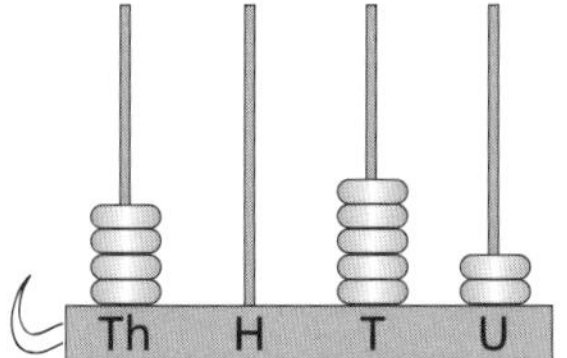

4. I have $3028 in the bank. How many $10 notes could I take out? ____
5. Measure the length of:
   a ▬▬▬▬ ____ cm ____ mm
   b ▬▬▬▬▬▬ ____ cm ____ mm
6. 1 tens and 6 ones
   + 2 tens and 2 ones
7. 45 bananas and 33 apples.
   How many more bananas? ____
8. 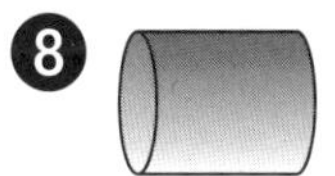 This is a ____________.
   Its bases are ____________.
   Its cross-section is a ____________.
9. a 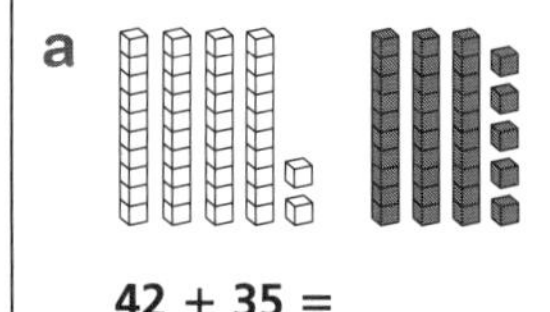 **42 + 35 =** ____
   b 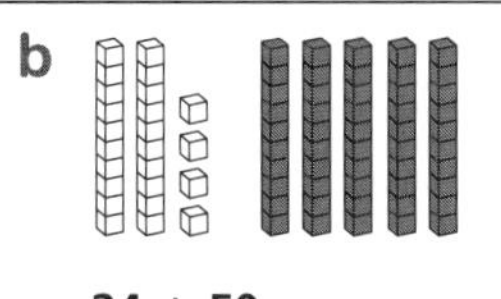 **24 + 50 =** ____
10. 100 more than 5376. ____

28:4 ☐ out of 5

Extension

1. The bottom part is cut off.
   The part left would be:
   A half of the volume
   B more than half
   C less than half ____
2. **a** How many centimetres in 1 m? ____
   b How many millimetres in 10 cm? ____
   c How many millimetres in 50 cm? ____
   d How many millimetres in 1 m? ____
3. The time is 4:35.
   How many more minutes until 5:00? ____
4. How many faces could be seen if I walk around this 3D object when it is placed on a table? ____

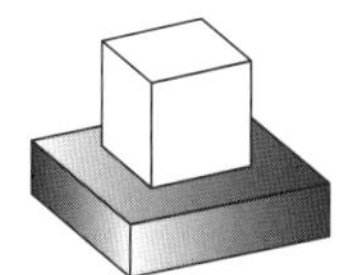

5. I am making cubes with toothpicks and Blu Tack. How many cubes could I make with 38 toothpicks? ____

Challenge

Draw a pyramid and describe it.

÷ tables

Groups of 2

÷ 2: 12, 2, 18, 10, 16, 0, 14, 8, 20, 6

Groups of 4

÷ 4: 4, 28, 20, 8, 32, 24, 40, 16, 36, 12

28 ÷ 4 = ☐
What number times 4 gives 28?

☐ × 4 = 28

29:1

out of 15

1. 35 + 20 _____
2. 28 + 41 _____
3. 56 − 11 _____
4. 68 − 22 _____
5. $\begin{array}{r} 29 \\ +\ \ 2 \\ \hline \end{array}$
6. 64 subtract 4. _____
7. 5 × 3 _____
8. Half of 14. _____
9. 6 × 2 _____
10. $\begin{array}{r} 69 \\ +\ \ 1 \\ \hline \end{array}$
11. 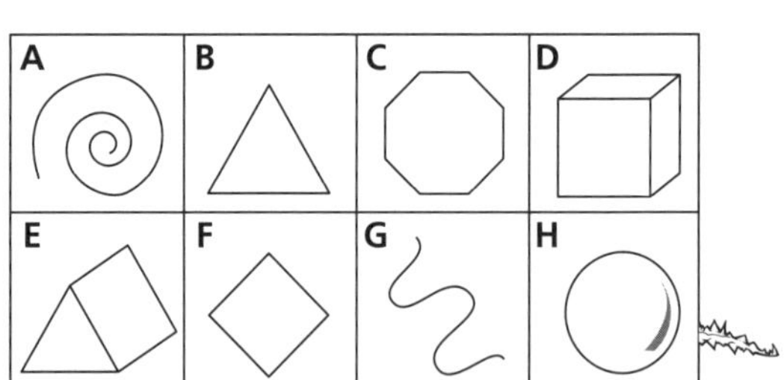

   a How many angles on shape **C**? _____

   b How many faces on shape **D**? _____

   c Name shape **E**. _____

   d Which are 2D shapes? _____

   e Which are solid objects? _____
12. These are called t__________ marks. 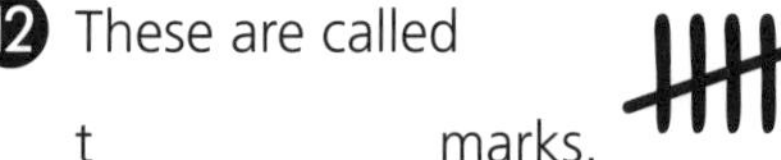
13. How many birds were there if there were 28 budgies and 7 quails?

   _____ + _____ = _____
14. 2000 + 400 + 20 + 1 = _____
15. Is 5270 an even number? _____

29:2

out of 16

1. 45 + 19 _____
2. 62 + 19 _____
3. 74 − 24 _____
4. 83 − 31 _____
5. $\begin{array}{r} 77 \\ +16 \\ \hline \end{array}$
6. 4 × 10 _____
7. 5 × 3 _____
8. 5 × 5 _____
9. 6 × 3 _____
10. $\begin{array}{r} 49 \\ +57 \\ \hline \end{array}$
11. Draw a column graph for this tally.

A	𝍸 II
B	IIII
C	𝍸 I

A								
B								
C								

12. This is a __________.

   It has _____ faces, _____ edges

   and _____ corners.

   The cross-section is a __________.
13. There were 46 seagulls and 36 magpies.

   How many birds altogether?

   _____ + _____ = _____
14. a How many faces has this pyramid? _____

   b How many corners has this pyramid? _____ 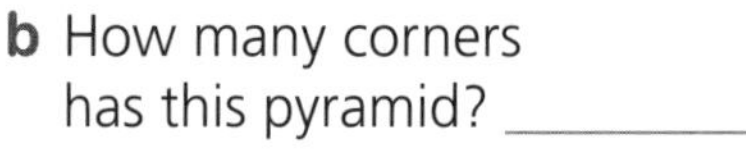
15. 6043 = ___ thousands + ___ tens + ___ ones 
16. 20 balls shared by 2 kids. How many each? _____

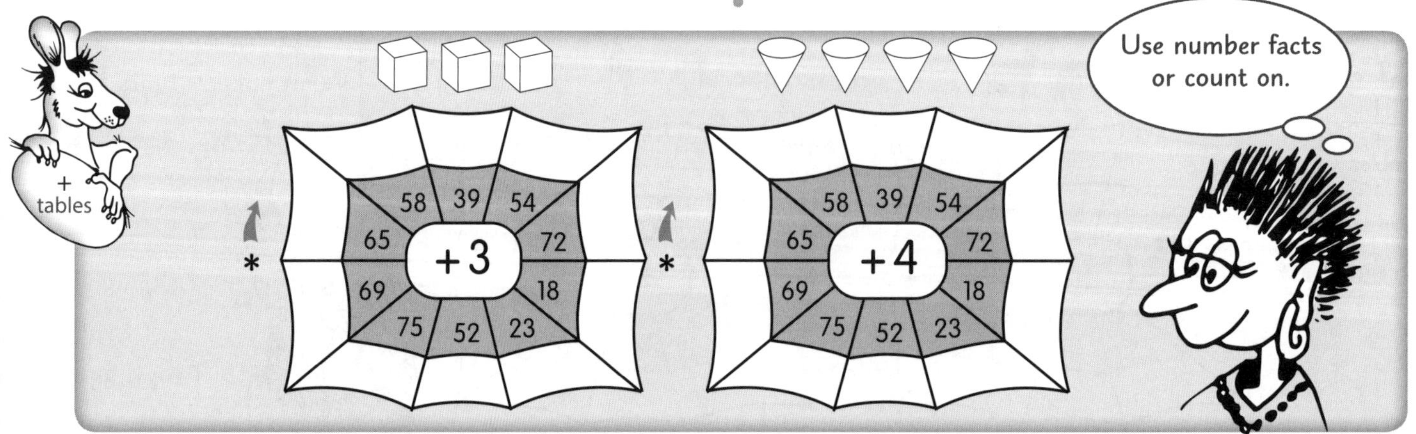

29:3 ☐ out of 6

1.

tens	ones
3	8
+	5

2.

tens	ones
5	7
+	5

3. There were 59 rosellas and 46 kookaburras.
How many birds altogether?
_____ + _____ = _____

4. a How wide is the graph below? _____

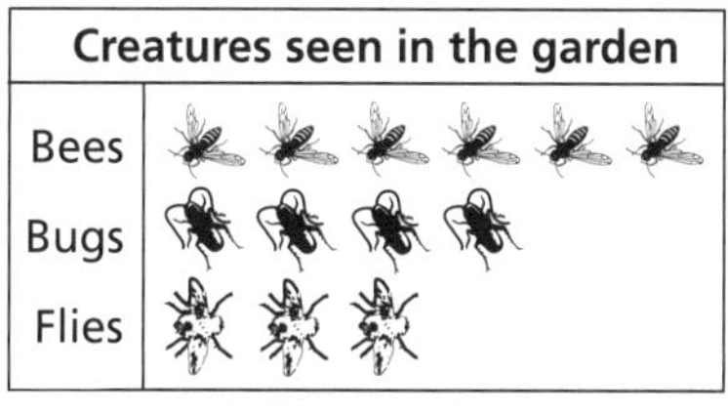

Each picture = 2 creatures

b How many more bees than flies were seen? _____

c How many creatures were seen altogether? _____

5. This is a _____________.
It has _____ faces,
_____ edges
and _____ corners.
The cross-section shape is a _____________.

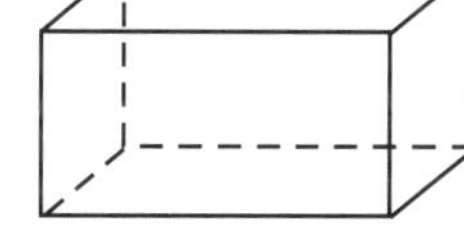

6. Name the two solids in this model. _____________
____

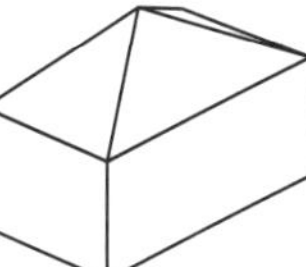

29:4 ☐ out of 4 — Extension

1. Guess my 3D object.
It has 5 faces. It has 8 edges.
The cross-section is a square.
____

Draw the object.

2. I have 13 fewer cards than my friend.
They have 52 cards.
How many cards do we have altogether?
_____ + _____ = _____

3. 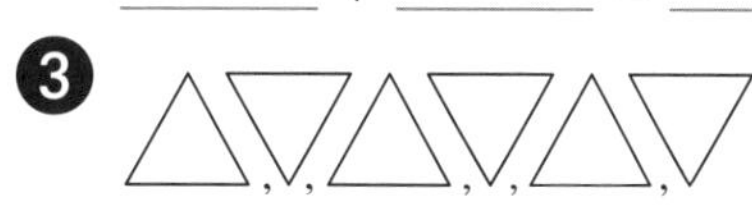

If this pattern continues, what will the 26th triangle look like? _____

4. Mum grew 36 roses and cut 18.
How many were not cut? _____

Challenge

Draw a prism and describe it.
____
____
____

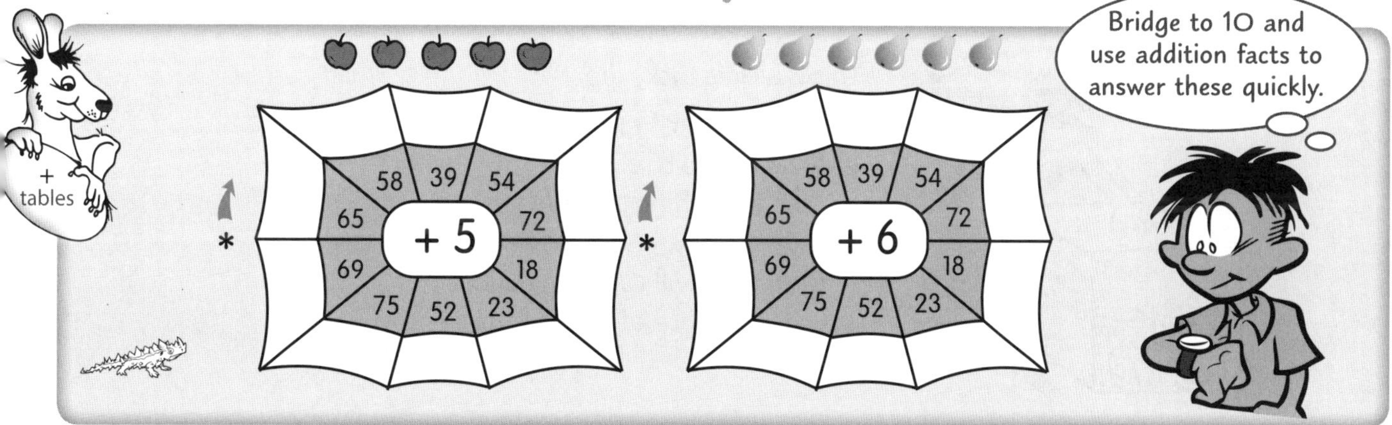

30:1

out of 16

1. 15 + _____ = 20
2. 27 + _____ = 30
3. 26 + _____ = 30
4. 39 + _____ = 40
5. 388 + 31 = _____
6. 60 plus 12 _____
7. 2 × 3 _____
8. 10 × 5 _____
9. 5 × 5 _____
10. 437 + 82 = _____
11. Write the number shown by:
 a 𝍸 𝍸 _____ b 𝍸 𝍸 𝍸 _____
12. I saw 29 kangaroos and 15 wallabies.
 How many did I see altogether?
 _____ + _____ = _____
13. If I toss a coin in the air, would heads be more likely than tails? _____

14. On this column graph show 50 trucks and 30 cars.

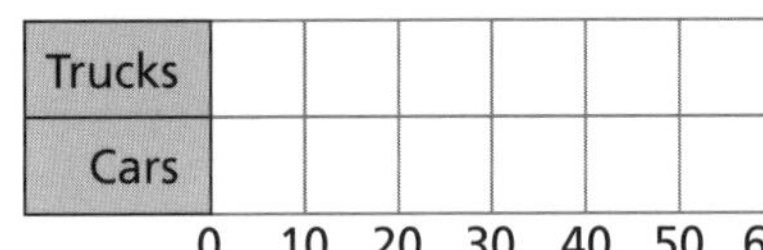

Trucks						
Cars						
	0 10	20	30	40	50	60

15. 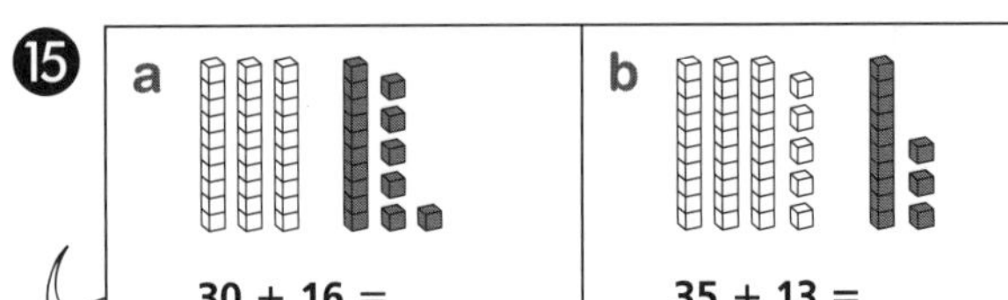
 a 30 + 16 = ___
 b 35 + 13 = ___
16. 16 shared between 4. How many each? _____
 △△△△△△△△△△△△△△△△

30:2

out of 14

1. 34 + _____ = 42
2. 28 + _____ = 36
3. 56 + _____ = 67
4. 79 + _____ = 85
5. 296 + 132 = _____
6. 7 × 3 _____
7. 8 × 4 _____
8. 9 × 2 _____
9. 5 × 4 _____
10. 504 + 459 = _____
11. impossible | unlikely | even chance | very likely | certain

 I throw a standard die once. Choose a label for the chance that the die shows:
 a a three _____
 b an odd number _____
 c larger than one _____
 d a number _____
12. I saw 36 snakes and 38 lizards.
 How many did I see altogether?
 _____ + _____ = _____
13. Round 453 to the nearest hundred. _____

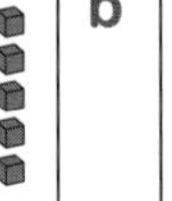

14. 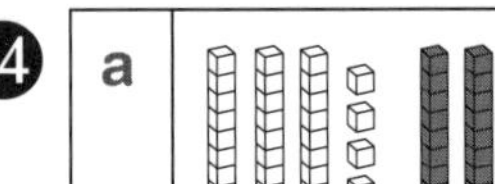 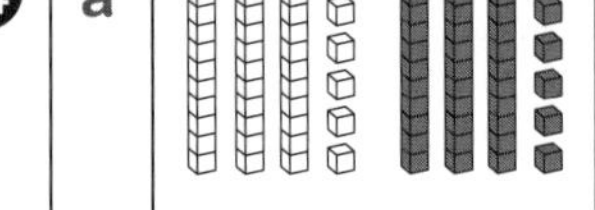 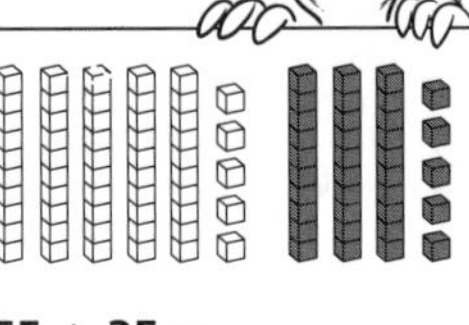
 a 35 + 35 = ___
 b 55 + 35 = ___

Chance

This is a class lucky dip.

a Is Gino likely to pick the pen? _____

b What is the chance that Gino will pick something starting with 'c'? _____

c If we add another different prize to the lucky dip, will Gino have more or less chance of picking the pen? _____

 ISBN 978 0 6557 0883 4

30:3  out of 6

1

hund	tens	ones
1	2	5
+ 4	3	7

2

hund	tens	ones
2	0	8
+ 1	9	3

3 Is each shape a prism, a pyramid or neither?

a 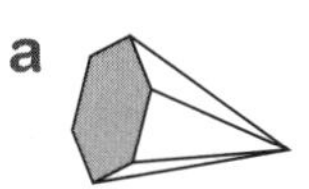 ______

b 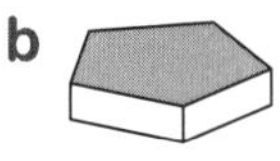 ______

4 I saw 38 wombats and 16 echidnas.

How many did I see altogether? ______

5 **Flowers picked**

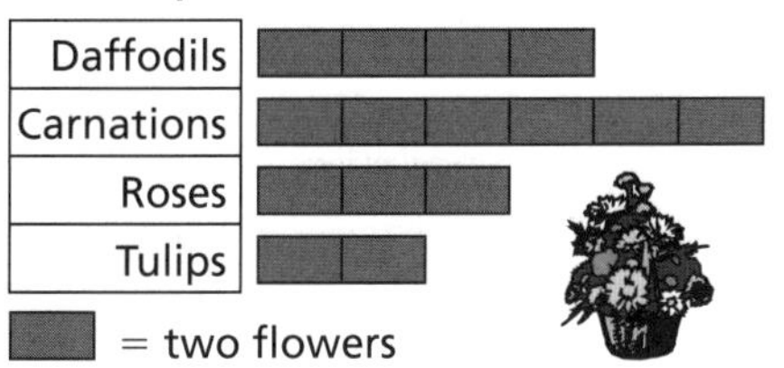

a The most popular flower. ______

b The least popular flower. ______

c How many daffodils? ______

d How many more carnations were picked than tulips? ______

e How many flowers altogether? ______

6

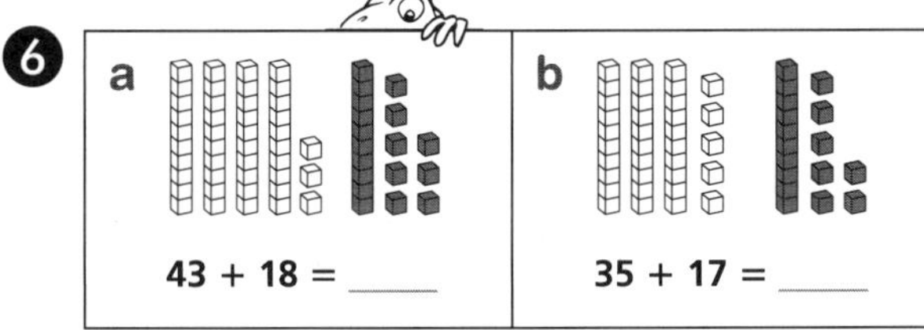

30:4  out of 4

Extension

1 Guess my 3D object.

It has 6 faces. It has 12 edges.

The cross-section is a square.

______

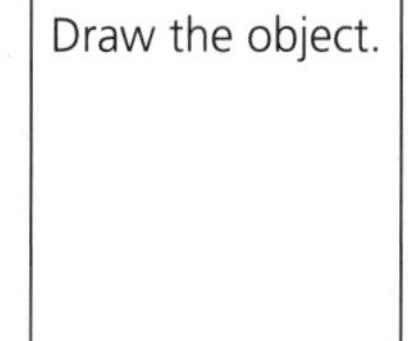

2 I have 17 fewer stickers than my friend.

She has 35 stickers.

How many stickers do we have altogether?

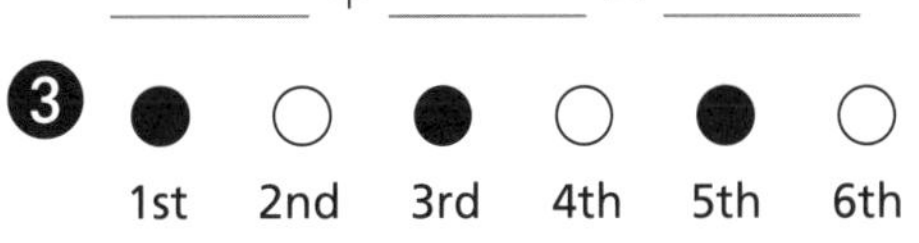

3 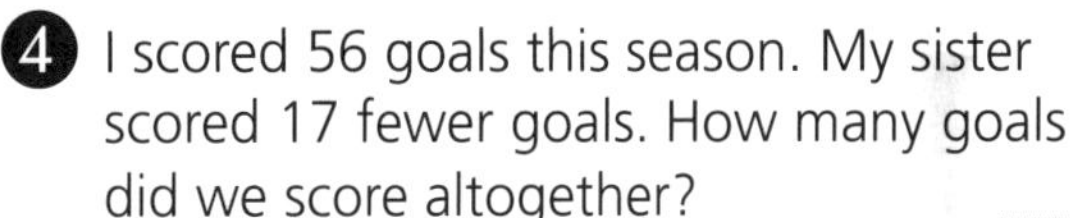

If this pattern continues, will the 41st circle be shaded? ______

4 I scored 56 goals this season. My sister scored 17 fewer goals. How many goals did we score altogether? ______

Challenge

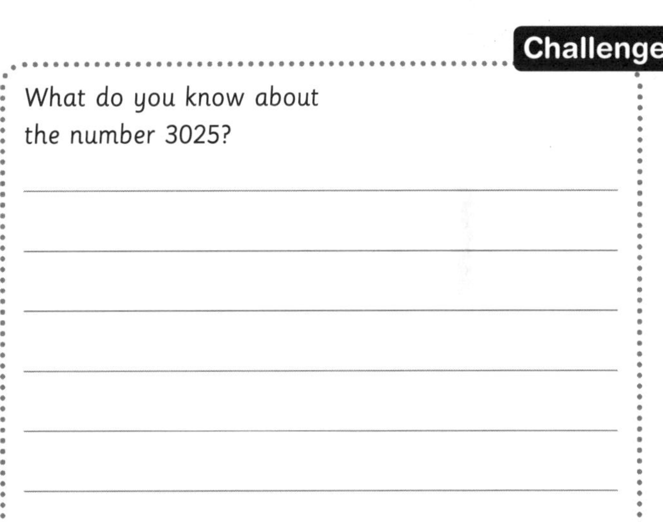

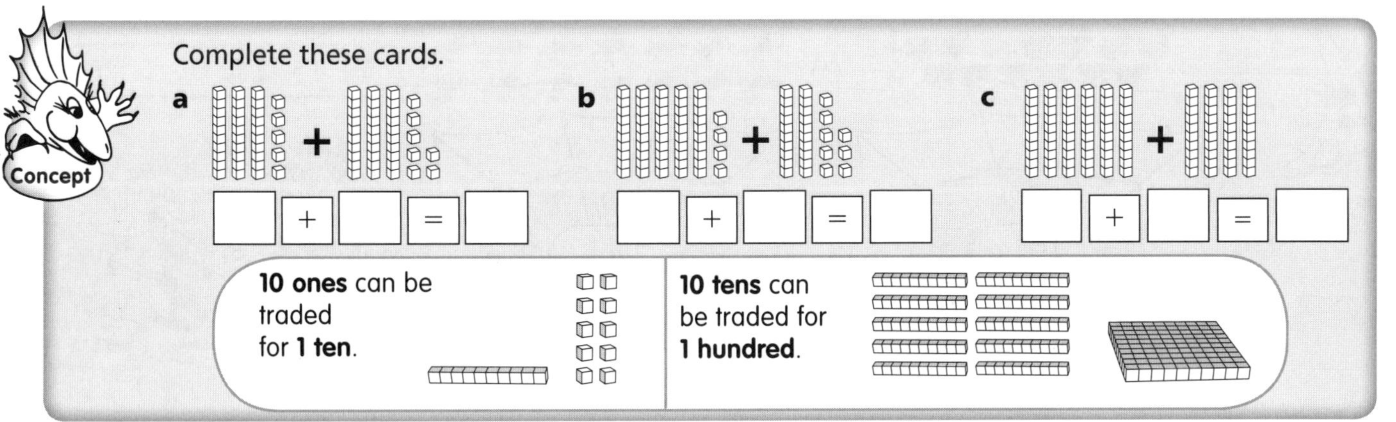

31:1 — out of 16

1. 19 + 20 _____
2. 30 + 21 _____
3. 61 − 12 _____
4. 82 − 13 _____
5. 572 + 53
6. 67 minus 14. _____
7. 8 × 2 _____
8. 5 × 3 _____
9. 7 × 10 _____
10. 672 + 251

11. a 51, 61, 71, _____, _____, _____, _____
    b 45, 50, 55, _____, _____, _____, _____
    c 3, 6, 9, _____, _____, _____, _____
    d 430, 440, 450, _____, _____
12. A 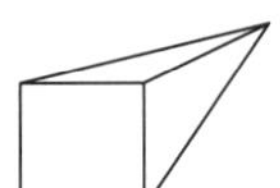 B  C 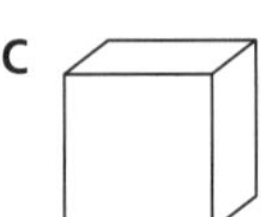
    Which shape is a pyramid? _____
13. Colour 4 fifths of this rectangle

14. Is 7205 or 7250 larger? _____
15. 
    a 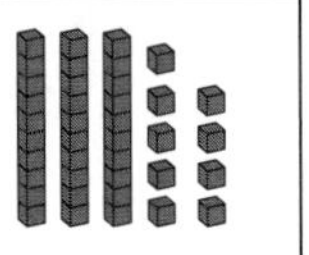 23 + 39 = _____

    b 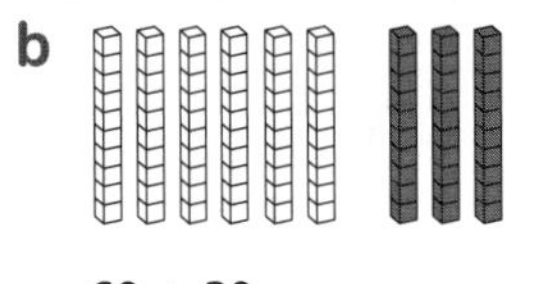 60 + 30 = _____
16. 10 + 6 = 16 so 16 − 6 = _____.

31:2 — out of 15

1. 300 + 45 _____
2. 120 + 41 _____
3. Half of 86. _____
4. 60 − 35 _____
5. 193 + 68
6. 62 minus 34. _____
7. 7 × 2 _____
8. 4 × 3 _____
9. 6 × 10 _____
10. 739 + 174

11. 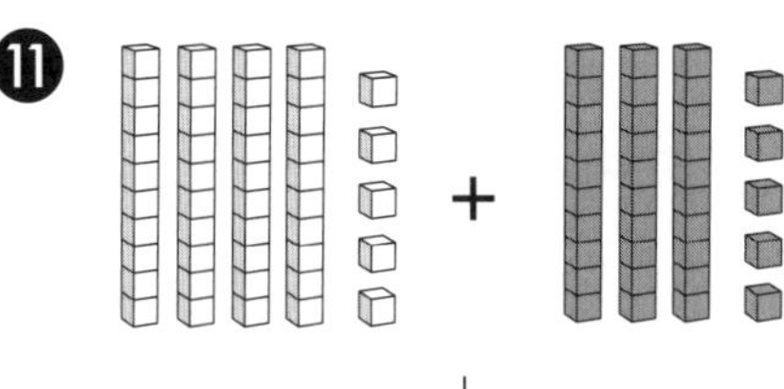
    _____ + _____ = _____
12. If I take a ball without looking:
    a am I more likely to pick a black or grey counter? _____
    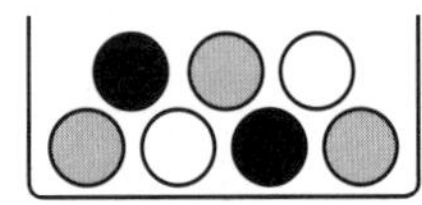
    b is there an equal chance of picking black, white and grey? _____
13. Jason drew 2 blue monsters.
    He gave each one 14 legs.
    How many legs were there altogether? _____
14. 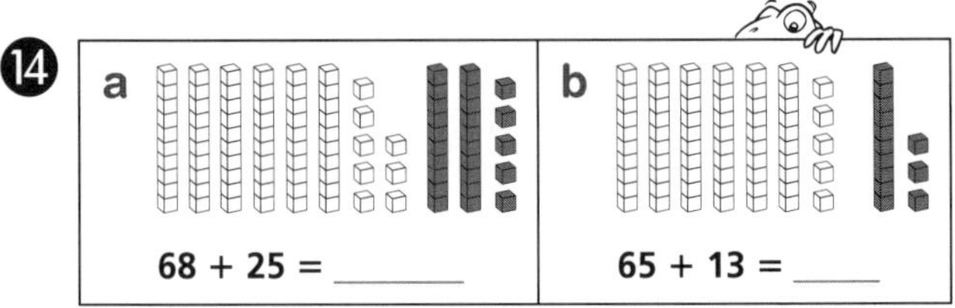
    a 68 + 25 = _____
    b 65 + 13 = _____
15. 18 shared by 2. How many each? _____

− tables

13 −	9	3	7	4	8	6	2	10	12	5
*										

14 −	9	0	7	4	8	6	2	10	12	5
*										

5 + 8 = 13 so 13 − 5 = 8.

 • *AUSTRALIAN SIGNPOST MATHS 3 MENTALS* • ISBN 978 0 6557 0883 4

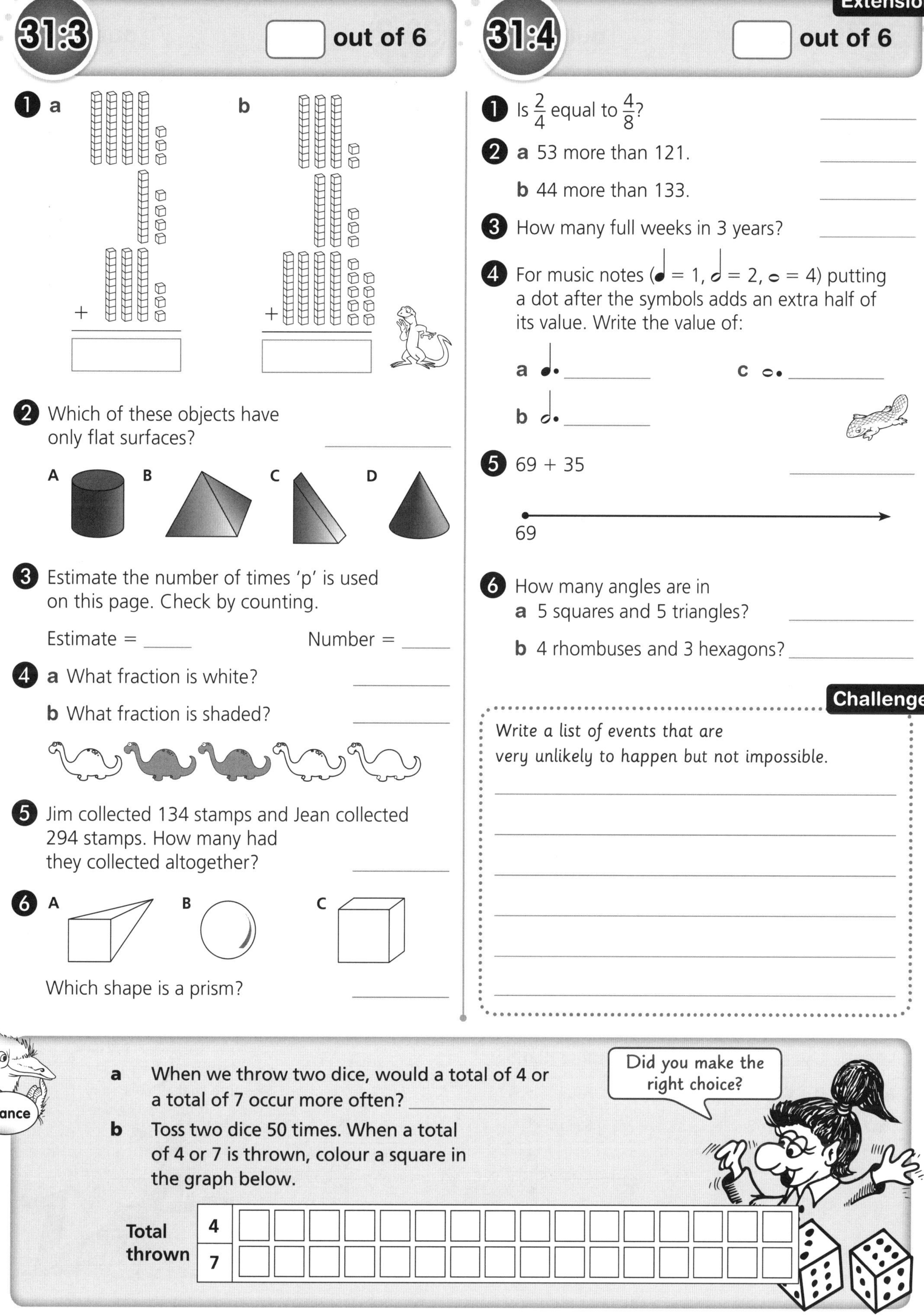

31:3 [ ] out of 6

1 a (base-ten blocks addition) + [ ]
b (base-ten blocks addition) + [ ]

2 Which of these objects have only flat surfaces? ________

A B C D

3 Estimate the number of times 'p' is used on this page. Check by counting.

Estimate = ____ Number = ____

4 a What fraction is white? ________

b What fraction is shaded? ________

5 Jim collected 134 stamps and Jean collected 294 stamps. How many had they collected altogether? ________

6 A B C

Which shape is a prism? ________

31:4 [ ] out of 6

Extension

1 Is $\frac{2}{4}$ equal to $\frac{4}{8}$? ________

2 a 53 more than 121. ________

b 44 more than 133. ________

3 How many full weeks in 3 years? ________

4 For music notes (♩ = 1, 𝅗𝅥 = 2, 𝅝 = 4) putting a dot after the symbols adds an extra half of its value. Write the value of:

a ♩. ________ c 𝅝. ________

b 𝅗𝅥. ________

5 69 + 35 ________

69

6 How many angles are in

a 5 squares and 5 triangles? ________

b 4 rhombuses and 3 hexagons? ________

Challenge

Write a list of events that are very unlikely to happen but not impossible.

____

Chance

a When we throw two dice, would a total of 4 or a total of 7 occur more often? ________

b Toss two dice 50 times. When a total of 4 or 7 is thrown, colour a square in the graph below.

Total thrown																	
	4																
	7																

32:1

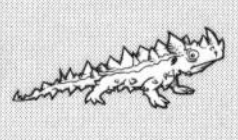

out of 15

1. 33 + 45 ____
2. 29 + 30 ____
3. 40 − 13 ____
4. 80 − 11 ____
5. 70 − 6 = ____
6. 67 minus 14. ____
7. 30 less than 55. ____
8. 13 subtract 5. ____
9. 7 groups of 10. ____
10. 83 − 7 = ____
11. I ran 100 metres 3 times. How far did I run? ____
12. 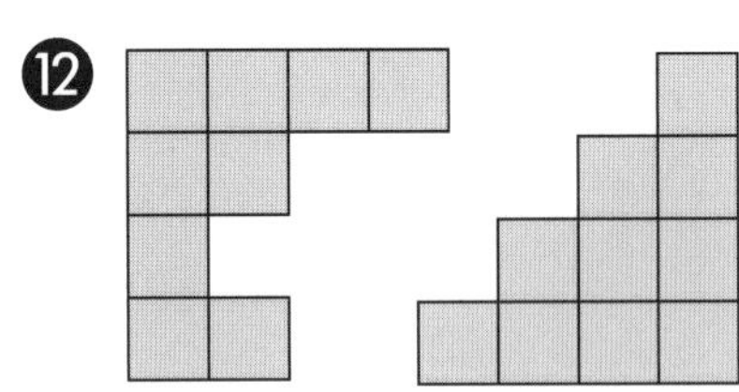
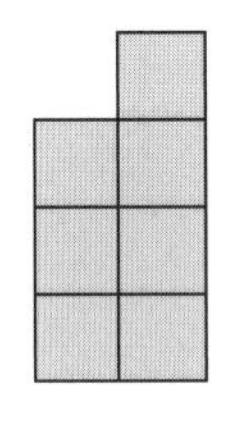

    a Which shape has the greatest area? ____

    b Which shape has the least area? ____

    c What is the total area of the shapes **A**, **B** and **C**? ____ square units
13. 3000 + 500 + 20 + 1 = ____
14. a 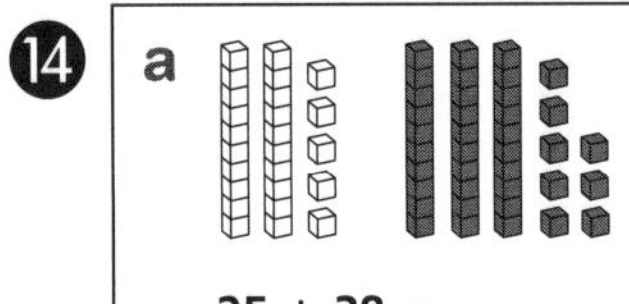
25 + 38 = ____

    b 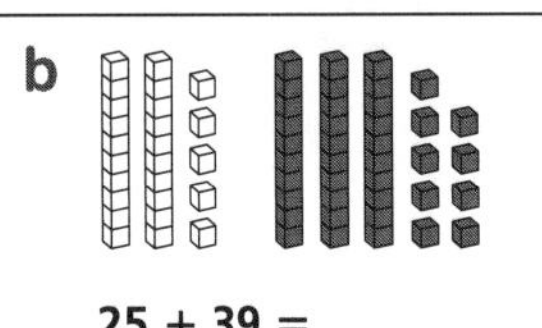
25 + 39 = ____
15. 15, 20, 25, ____, ____, ____, ____, ____

32:2

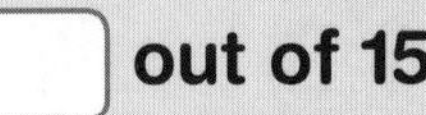

out of 15

1. 400 + 29 ____
2. 140 + 32 ____
3. Half of 62. ____
4. 40 − 15 ____
5. 90 − 37 = ____
6. 4 × 4 ____
7. 7 × 2 ____
8. 4 × 3 ____
9. 6 × 10 ____
10. 47 − 8 = ____
11. I walked along the balance beam 5 times. If it is 5 metres long, how far did I walk? ____
12. a This time is read as ____.

    b It means ____ minutes past ____.

13. This area is 3 rows of ____. Area = ____ cm^2

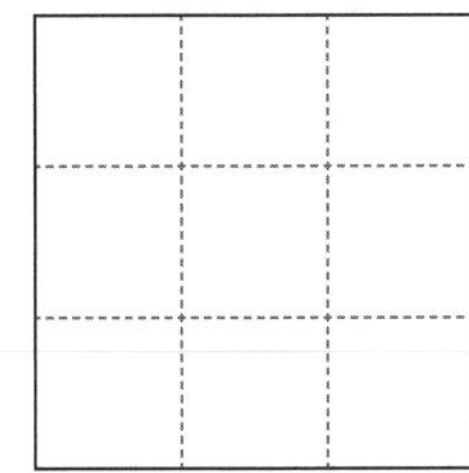

14. 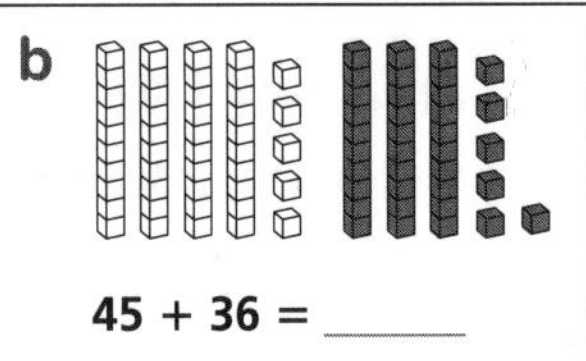
a **25 + 36 =** ____
b **45 + 36 =** ____
15. 9 shared by 3. How many each? ____

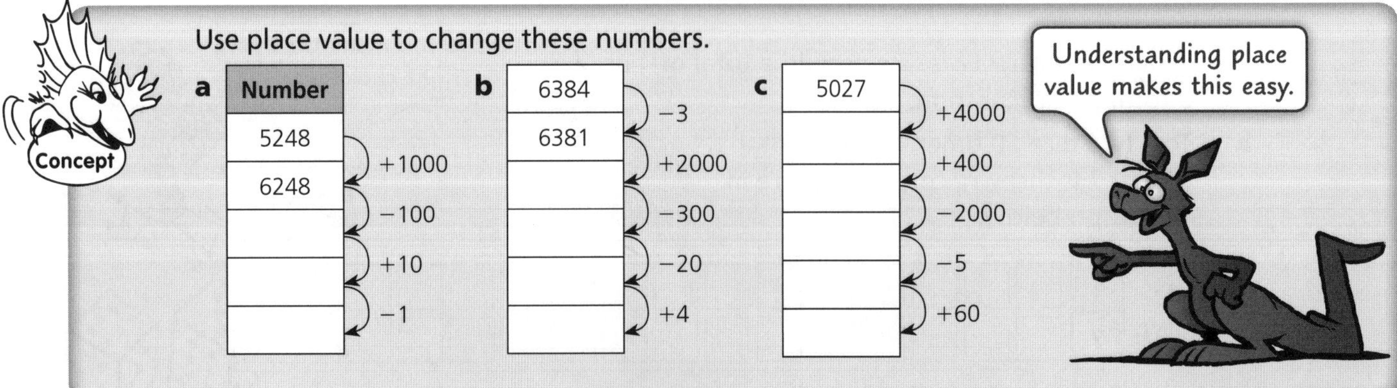

Use place value to change these numbers.

a

Number	
5248	+1000
6248	−100
	+10
	−1

b

6384	−3
6381	+2000
	−300
	−20
	+4

c

5027	+4000
	+400
	−2000
	−5
	+60

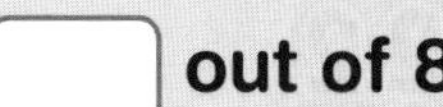

32:3 ☐ out of 8

❶ 40 − 7 = ____

❷ 60 − 9 = ____

❸ 80 − 3 = ____

❹ This is a ______________________.

It has _____ faces, _____ edges

and _____ corners.

The cross-section is a ________________.

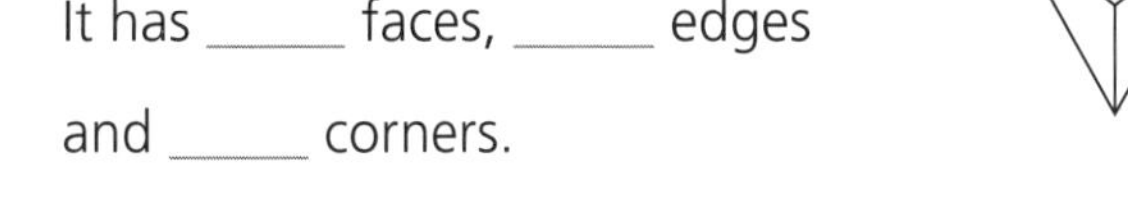

❺ 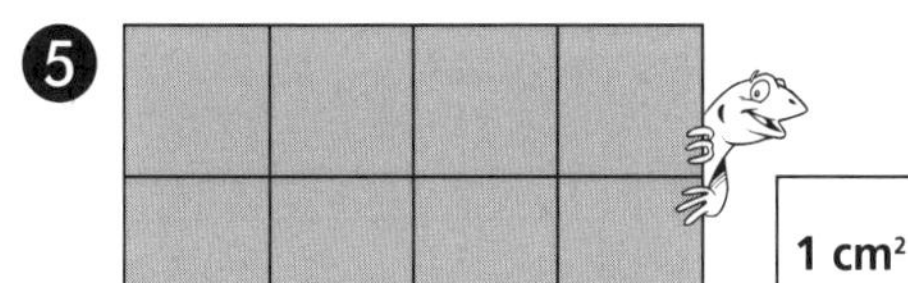 

The grey area is 2 rows of _____.

Area = _____cm²

❻ Complete the labels.

a 

4: ☐☐

☐ past 4

b 

10: ☐☐

☐ past 10

❼ **a** Is 3405 higher than 3412? __________

b Is 5089 higher than 5800? __________

❽ **a** 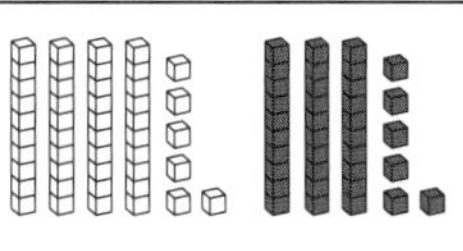

46 + 36 = ______

b 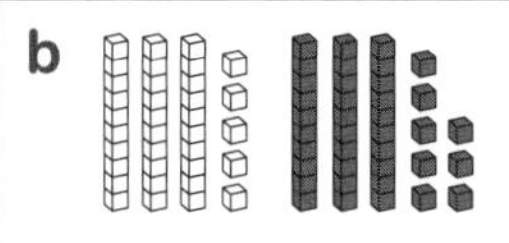

35 + 38 = ______

32:4 ☐ out of 4

Extension

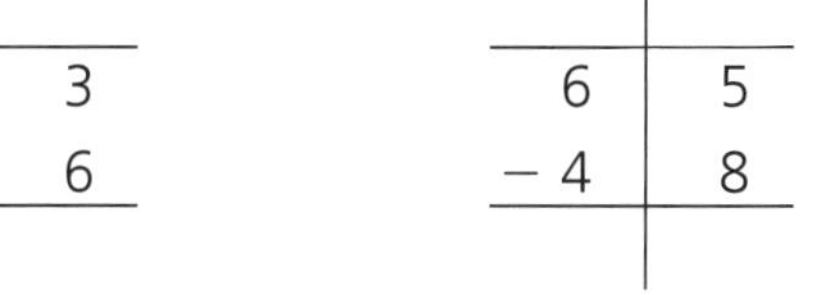

❶

tens	ones
4	3
− 1	6

❷

tens	ones
6	5
− 4	8

❸ **A** 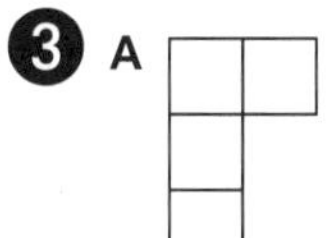 **B** 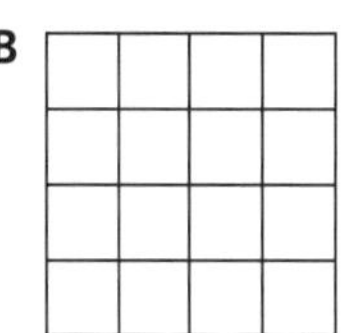

How many of shape **A** could be cut from shape **B**? __________

❹ **a** 1000 – 304 _____ **b** 1000 – 209 _____

c 1000 – 701 _____ **d** 1000 – 202 _____

Challenge

This area has 12 squares. Draw and label different rectangles that also have an area of 12 squares.

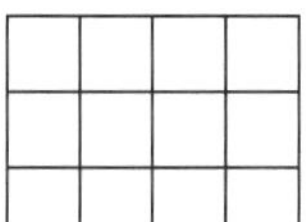

3 rows of 4 = 12

33:1 ☐ out of 17

1. 20 + 13 _____
2. 30 + 15 _____
3. 20 − 17 _____
4. 20 − 14 _____
5. $\begin{array}{r} 42 \\ -\,28 \\ \hline \end{array}$
6. 54 minus 12. _____
7. 9 × 2 _____
8. 3 × 3 _____
9. 9 × 10 _____
10. $\begin{array}{r} 53 \\ -\,25 \\ \hline \end{array}$
11. The length of my foot is 20 cm. If the length of my chair was four lengths of my feet, how long is my chair? _____
12. This area is 2 rows of _____. Area = _____ cm^2

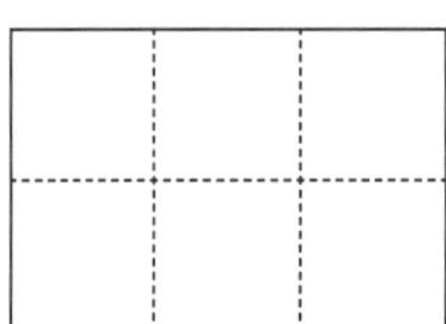

13. Is addition the opposite of subtraction? _____
14. List these angles in order of size, from smallest to largest. _____

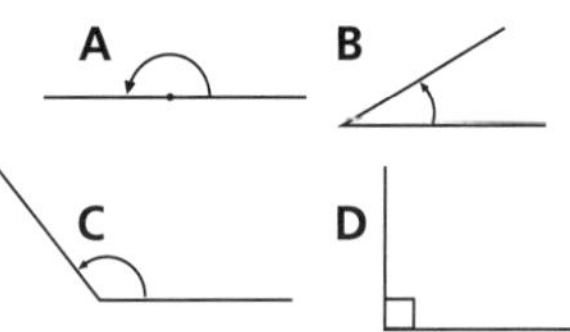

15. Surveys are used to collect _____.
16. Is 4609 lower than 4099? _____
17. 14 cars shared by 2. How many each? _____

33:2 ☐ out of 14

1. 56 + 23 _____
2. 16 + 16 _____
3. 40 − 13 _____
4. 80 − 11 _____
5. $\begin{array}{r} 50 \\ -\,15 \\ \hline \end{array}$
6. Double 18. _____
7. 5 groups of 3. _____
8. 6 rows of 5. _____
9. 6 rows of 10. _____
10. $\begin{array}{r} 61 \\ -\,36 \\ \hline \end{array}$

11. My ribbon is 68 cm. If I cut it in half, how long will each piece be? _____
12. This area is 3 rows of _____. Area = _____ cm^2

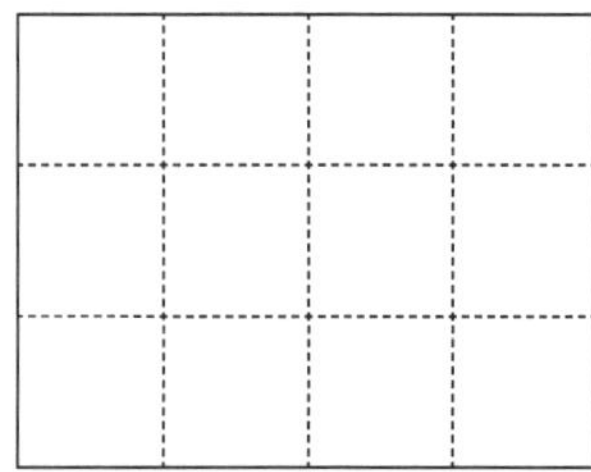

13. **a** 45 + 36 = _____ so 81 − 36 = _____
    b 27 + 38 = _____ so 65 − 38 = _____
    c 39 + 26 = _____ so 65 − 26 = _____
14. **impossible** **unlikely** **even chance** **likely** **certain**

I throw a standard die once. Choose a label for the chance that the die shows:

a an even number _____
b a seven _____
c smaller than two _____
d 1, 2, 3, 4, 5 or 6 _____

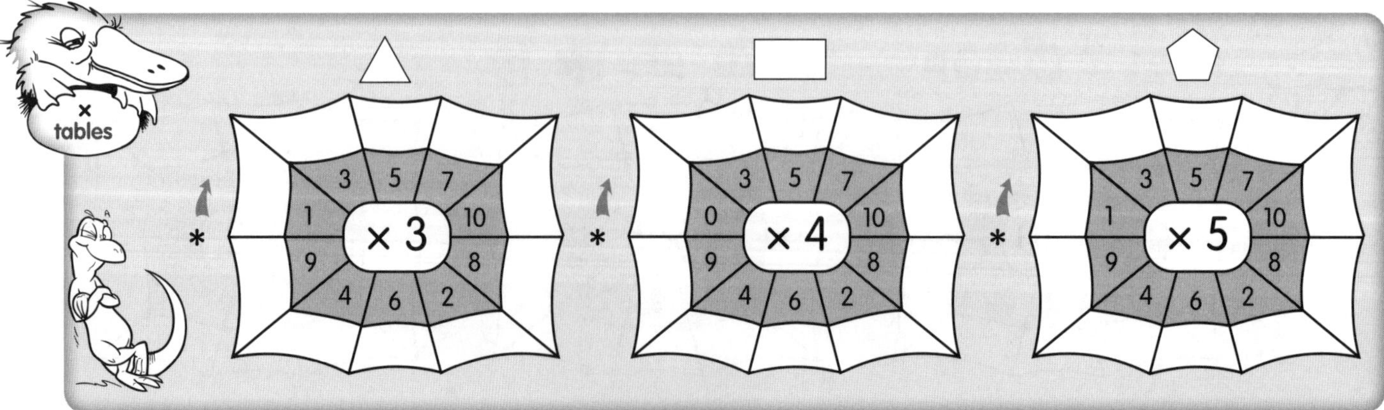

33:3 ☐ out of 10

1. 42 − 4 = ____

2. 51 − 3 = ____

3. 73 − 5 = ____

4.

tens	ones
$4	4
+$5	3

5.

tens	ones
$6	3
+$2	1

6. 52 + 29 = ____ so 81 − 29 = ____

7. 

Goals kicked this year

Kym	𝍸 𝍸 𝍸 𝍸
Lisa	𝍸 ||||
Diane	𝍸 𝍸 |||
Marta	𝍸 𝍸 ||

a Who kicked 13 goals? ____

b How many did Marta kick? ____

c How many more goals did Kym kick than Lisa? ____

8. This line is ____ centimetres long.

This line is ____ millimetres long.

9. Is 5091 larger than 5901? ____

10. **a** 52 + 26 = ____ so 78 − 26 = ____

b 35 + 46 = ____ so 81 − 46 = ____

33:4 Extension ☐ out of 4

1. I placed 6 square tiles in a long line with no gaps or overlaps. If each tile was 15 cm, how far was my line of tiles? ____

2. How many blocks are needed to make a staircase of ten steps? ____

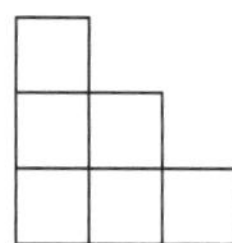

3. Use the sign to find the distance from:

a Minden to Ironbark ____

b Minden to Goodna ____

c Ironbark to Goodna ____

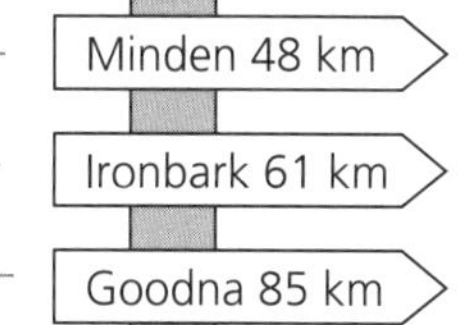

4. Which has the greatest area? ____

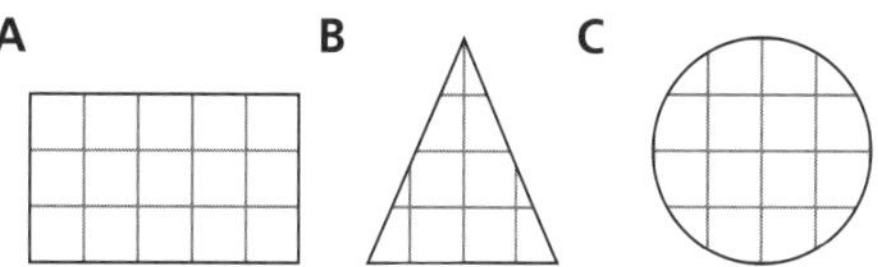

Challenge

Draw and label some fractions of your own.

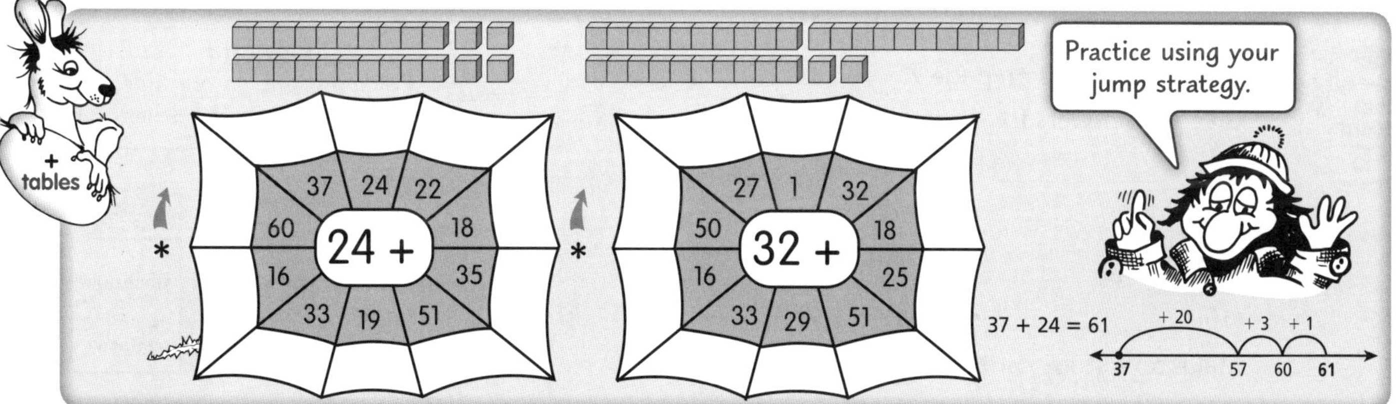

34:1 ☐ out of 13

1. 15 + ____ = 20
2. 28 + ____ = 30
3. 13 + ____ = 20
4. 14 + ____ = 20
5. $\begin{array}{r} 41 \\ -\ 29 \\ \hline \end{array}$
6. Double 12. ____
7. Halve 20. ____
8. 4×2 ____
9. 3×5 ____
10. $\begin{array}{r} 36 \\ -\ 18 \\ \hline \end{array}$

11. Write these decimals as tenths.

a 0·4 $\frac{\square}{10}$ **b** 0·7 $\frac{\square}{\square}$ **c** 0·9 $\frac{\square}{\square}$

12.

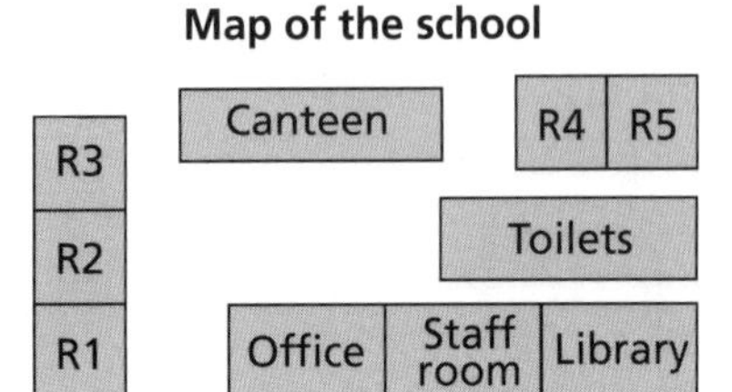

a What rooms are next to R2? ____

b What room is next to the office? ____

c What is between R3 and R4? ____

d What rooms are next to the staff room? ____

e What is between R5 and the library? ____

13. **a** 23, 33, 43, ____, ____, ____, ____

b 80, 75, 70, ____, ____, ____, ____

34:2 ☐ out of 14

1. 18 + ____ = 32
2. 21 + ____ = 36
3. 16 + ____ = 28
4. 42 + ____ = 59
5. $\begin{array}{r} 52 \\ -\ 37 \\ \hline \end{array}$
6. Double 19. ____
7. Halve 56. ____
8. 8×2 ____
9. 8×5 ____
10. $\begin{array}{r} 51 \\ -\ 25 \\ \hline \end{array}$

11.

a How many students came to school by bus or bike? ____

b How many more students walked to school than came to school by taxi? ____

12. My handspan is 15 cm. I measured my table and it was 4 handspans wide.

How wide is my table? ____

13. Write the decimal for:

a $\frac{3}{10}$ ____ **b** $\frac{5}{10}$ ____ **c** $\frac{8}{10}$ ____

d $1\frac{2}{10}$ ____ **e** $3\frac{7}{10}$ ____

14. **a** 45 + 25 = ____ so 70 − 25 = ____

b 24 + 29 = ____ so 53 − 29 = ____

Turn to ID card B on page 7.
Give the answers for these numbers.

(8) ____ (9) ____

(10) ____ (11) ____

(12) ____ (13) ____

(16) ____ shapes (17) ____ shapes

Make up a study card for any mistakes.

Study card
Put questions on one side and answers on the other.

34:3 out of 5

❶

tens	ones
4	6
− 2	8

❷

tens	ones
$7	5
−$4	9

❸

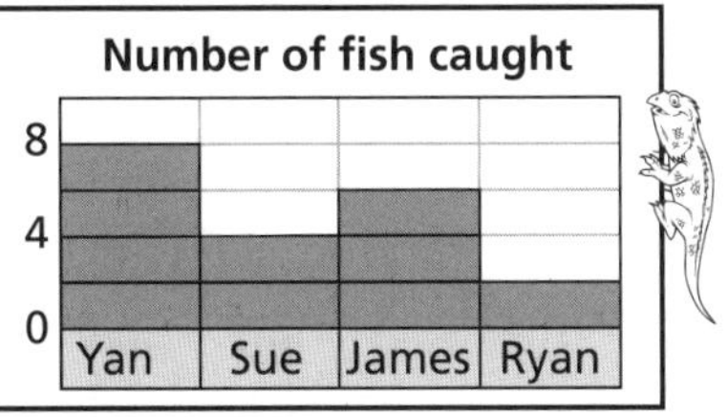

a Who caught the most fish? ________

b How many fish did Sue and Ryan catch altogether? ________

c How many more fish did Yan catch than Sue? ________

d How many fish were caught? ________

❹

Name	Tally	Total
Koala	卌 卌 卌 卌 II	
Bilby	卌 III	
Emu	卌 卌 卌 IIII	
Possum	卌 卌 卌 卌 卌 III	

a Fill in the last column of the tally above.

b Which animal was most common? ________

❺ Write the decimal for:

a 6 tenths ________ b 9 tenths ________

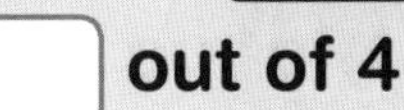

34:4 out of 4

❶

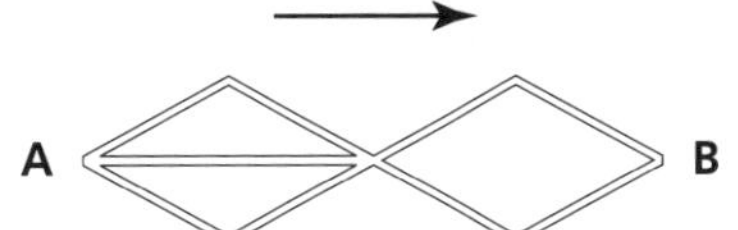

How many different ways are there of going from **A** to **B**? ________

❷ a 3 hours before 1:00.

____

b 47 minutes before 1:00.

____

❸ In how many different orders can you write the letters A, B and C? (You could make a list.) ________

❹ How many place-value ones blocks do you need to cover this area?

Area = ______ blocks

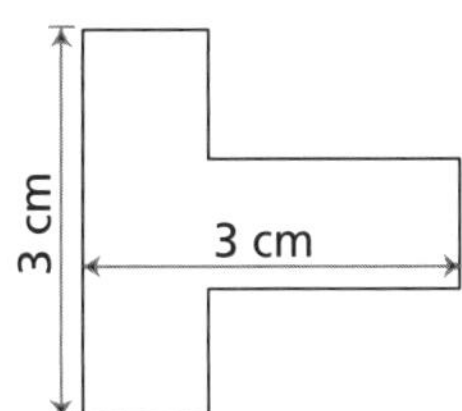

Challenge

Draw a path on the grid from the counter to the triangle. Describe the path.

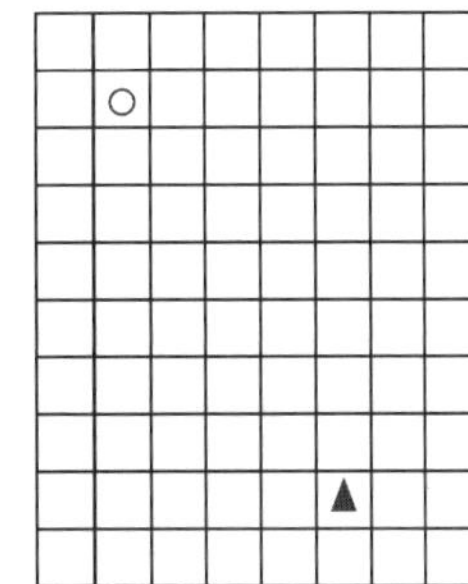

____

____

____

____

____

Use place value to change these numbers.

a

Number	
7027	
7028	+1
	−1000
	+100
	−10

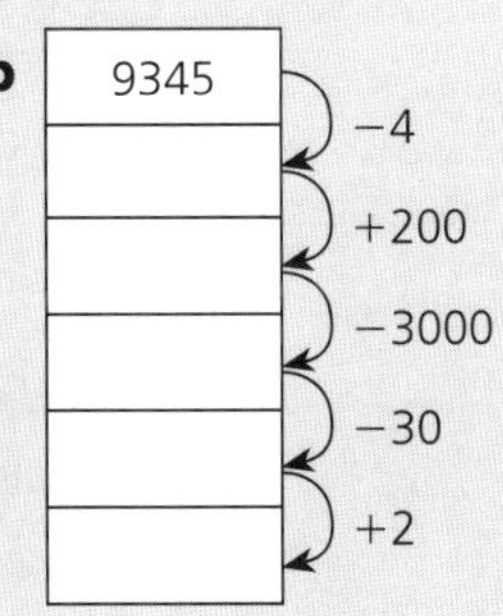

b

9345	
	−4
	+200
	−3000
	−30
	+2

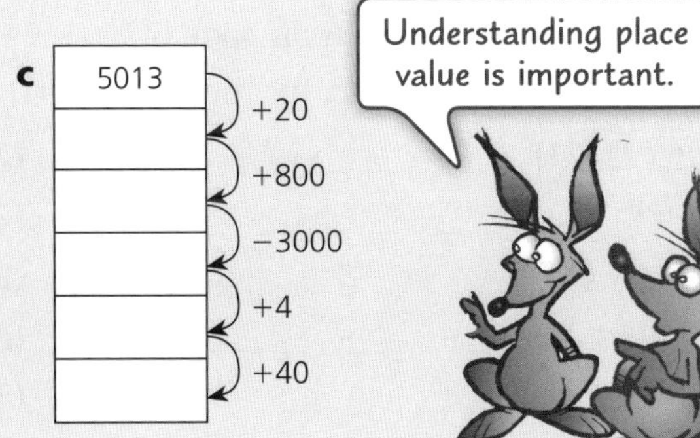

c

5013	
	+20
	+800
	−3000
	+4
	+40

35:1 out of 16

1. 8 × 2 ____
2. 4 × 5 ____
3. 3 × 10 ____
4. 6 × 5 ____
5. 62 − 13
6. Half of 16. ____
7. 4 less than 23. ____
8. 35 + 10 ____
9. 37 − 10 ____
10. 48 − 29
11. **a** 0·1, 0·2, 0·3, ____, ____, ____

    b $\frac{1}{10}$, $\frac{2}{10}$, $\frac{3}{10}$, $\frac{\square}{\square}$, $\frac{\square}{\square}$, $\frac{\square}{\square}$
12. Write the decimal for:

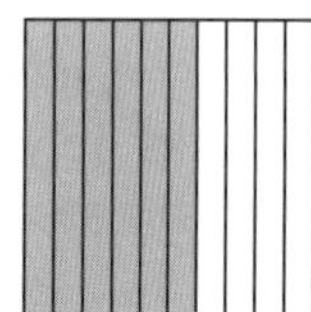

tens	ones	tenths
0		.

13. Use decimals to write zero point six. ____
14. 30 000 + 4000 + 300 + 20 + 1 = ____
15. Circle the largest number.

    456 701   465 780   456 710
16. Describe this 3D object.

____

____

35:2 out of 15

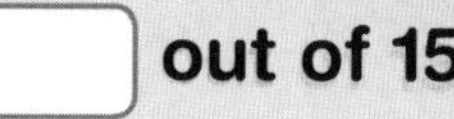

1. 7 × 2 ____
2. 8 × 5 ____
3. 6 × 10 ____
4. 7 × 5 ____
5. 51 − 28
6. Half of 62. ____
7. 9 less than 92. ____
8. 35 + 37 ____
9. 42 − 35 ____
10. 65 − 47
11. In this maze, Naomi started at A. Where did she end up if she turned left at the first corner, then right at the next, then left at the next?

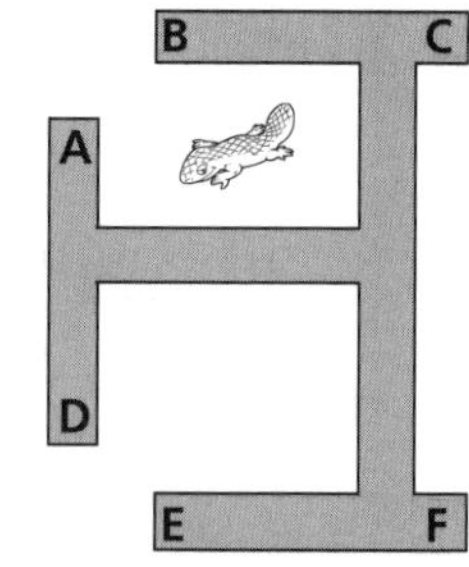

____

12. Write these decimals as tenths.

    a 0·3 $\frac{\square}{10}$ **b** 0·6 $\frac{\square}{\square}$ **c** 0·9 $\frac{\square}{\square}$
13. The fraction for 3.4 is $3\frac{4}{10}$.

    Write the fraction for:

    a 1·8 $\frac{\square}{\square}$ **b** 3·9 $\frac{\square}{\square}$ **c** 6·7 $\frac{\square}{\square}$
14. Circle the largest number.

    830 871,   830 781,   830 187
15. How many digits in 8 756 430? ____

Turn to ID card A on page 6.
Give the answers for these numbers.

(21) ____		(22) ____	
(23) ____	graph	(24) ____	
(25) ____	graph	(26) ____	graph
(27) ____	watch	(28) ____	clock
(29) ____		(30) ____	scales

Learn the terms you do not know.

 • *AUSTRALIAN SIGNPOST MATHS 3 MENTALS* • ISBN 978 0 6557 0883 4

35:3 ☐ out of 6

1. a 70 000 + 3000 + 700 + 40 + 1 = ________

   b 90 000 + 600 + 30 + 9 = ________

2. Write these decimals as tenths.

   a 0·2 $\frac{\square}{\square}$ b 0·8 $\frac{\square}{\square}$ c 0·1 $\frac{\square}{\square}$

3. Match each fraction to its decimal.

$\frac{5}{10}$	0·8
$\frac{8}{10}$	0·3
$\frac{3}{10}$	0·5

$1\frac{7}{10}$	1·9
$1\frac{2}{10}$	1·2
$1\frac{9}{10}$	1·7

4. Write the decimal for:

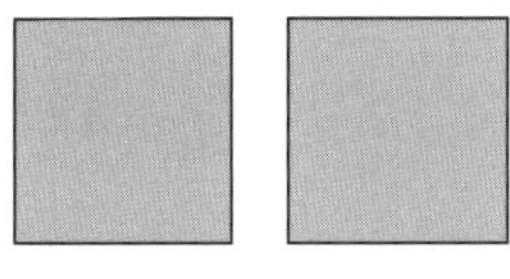

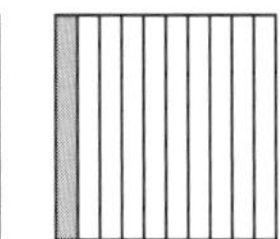

tens	ones	.	tenths
		.	

5. Write these numbers on the place-value chart.

   a 567 603 b 380 689 c 293 004

	Thousands	Hundreds	Tens	Units
a	567	6	0	3
b				
c				

6. Is 5 678 342 larger than 5 678 451? ________

35:4 ☐ out of 5

Extension

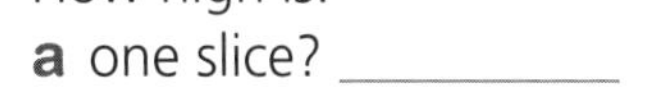
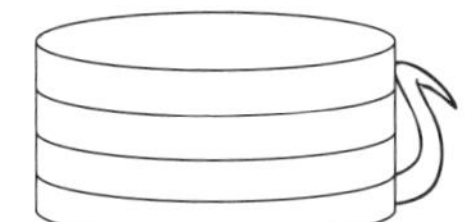

1. This cylinder is 8 cm high.
   How high is:
   a one slice? ________
   b three slices? ________

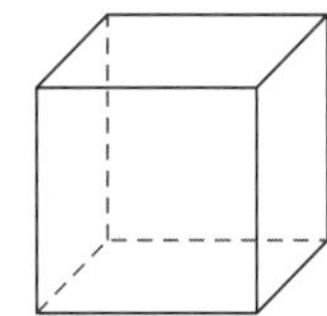

2. The number of faces plus the number of corners minus the number of edges.

   ____

3. Circle the largest number.

   6 560 936, 6 560 963, 6 506 937

4. 30 + 600 + 1 000 000 + 70 000 = ________

5. Write the numeral seven thousand and four. ________

Challenge

Draw a map of a room in your home.
Draw a key to show things in the room.

Turn to ID card B on page 7.
Give the answers for these numbers.

(20) net of a ________ (24) ________

(25) ________ (26) ________

(28) ________ (29) ________

(30) ________

Fill out the table for this cube.

	Faces	Edges	Corners
Number			

36:1 out of 15

1. 13 + 7 _____
2. 23 + 7 _____
3. 33 + 7 _____
4. 43 + 7 _____
5. 91 − 24 _____
6. 18 + _____ = 20
7. 17 + _____ = 20
8. 8 + _____ = 20
9. 7 + _____ = 20
10. 55 − 16 _____
11. Write these decimals as tenths.
    a 0·2 $\frac{}{10}$ b 0·4 $\frac{}{}$ c 0·8 $\frac{}{}$
12. Circle the largest number.
    5 091 657 5 901 657 5 091 567
13. Write the decimal for:

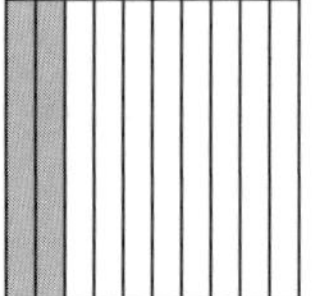

tens	ones	.	tenths
0		.	

14. What two shapes are used to make this model? _____ _____

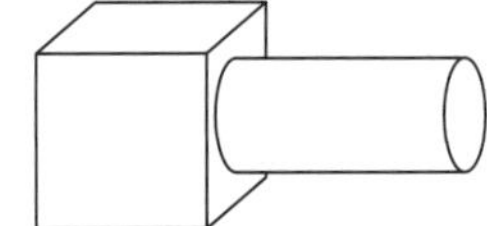

15. Bridge to the next ten to find:
    a 36 + 5 _____ b 28 + 5 _____
    c 57 + 6 _____ d 69 + 3 _____

36:2 out of 16

1. 36 + 9 _____
2. 136 + 9 _____
3. 43 − 19 _____
4. 57 − 19 _____
5. 78 − 49 _____

6. 755 − 7 _____
7. 462 − 4 _____
8. 65 + 19 _____
9. 45 + 19 _____
10. 31 − 18 _____
11. The fraction for 3·4 is $3\frac{4}{10}$
    Write the fraction for:
    a 2·3 b 4·5 c 7·2
12. Is this a pyramid, a prism or neither? _____

13. How many digits in 546 790? _____
14. Is this the net of a cube? _____

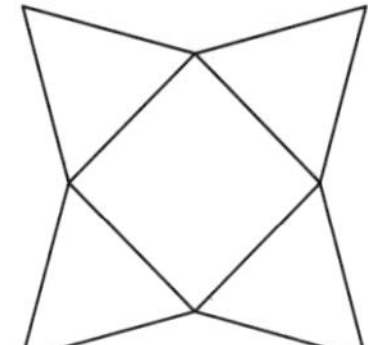

15. Use the jump strategy to find:
    67 + 28 = _____

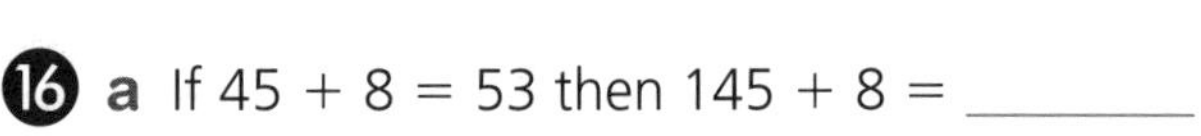

16. a If 45 + 8 = 53 then 145 + 8 = _____
    b If 35 + 7 = 42 then 135 + 7 = _____

– tables

15 –	8	5	12	7	10	3	4	9	6	11

16 –	8	5	11	7	10	3	12	9	6	2

16 − 2?

 • *AUSTRALIAN SIGNPOST MATHS 3 MENTALS* • ISBN 978 0 6557 0883 4

36:3 ☐ out of 6

1. Write the decimal for:

  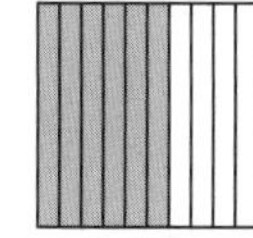

tens	ones	.	tenths
		.	

2. Write these decimals as tenths.

a 0·5 $\frac{\square}{\square}$ b 0·9 $\frac{\square}{\square}$ c 0·2 $\frac{\square}{\square}$

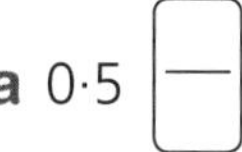 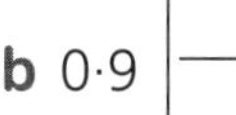

3. Is this the net of a cube? ______

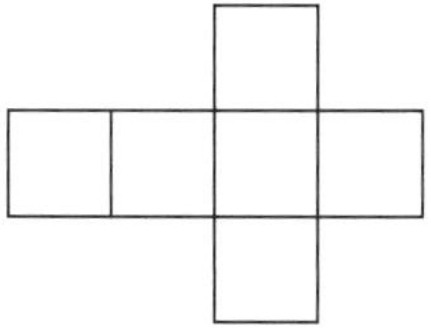

4. Bridge to the next ten to find:

a 49 + 8 ____ b 45 + 7 ____

c 89 + 9 ____ d 37 + 5 ____

5. Use the jump strategy to find:

a 49 + 36 = ____

b 72 − 34 = ____

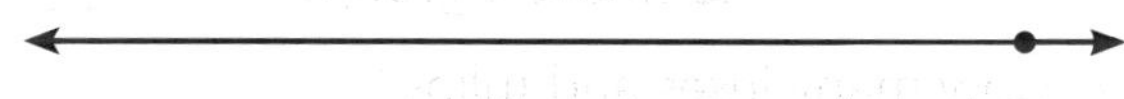

6. What is the cross-section of a:

a cylinder? ______

b cube? ______

c sphere? ______

36:4 ☐ out of 5

Extension

1. 6 + 500 000 + 6 000 000 = ______

2. If this pattern continues how many hexagons will be in row:

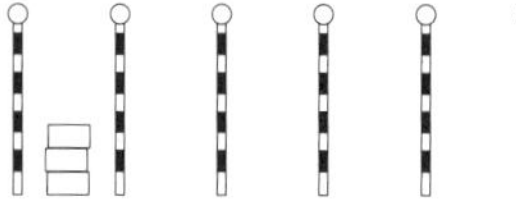

a 5? ______

b 10? ______ c 50? ______

3. Three bricks were placed between each pair of poles.

a How many bricks were used? ______

b How many would be used if there were 10 poles? ______

4.   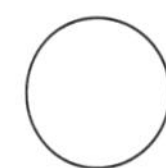  

If this pattern was repeated, what would the 19th shape be? ______

5. Luke has 12 books. Kim has 3 times as many. How many has Kim? ______

Challenge

Draw and describe a 3D object.

______
______
______
______
______
______
______

tables

−2: 5, 7, 10, 6, 8, 4, 9, 5, 11, 2

−4: 7, 14, 6, 10, 12, 8, 13, 9, 5, 11

even − even = ____
odd − even = ____

37:1 out of 14

1. 18 + 5 ______
2. 28 + 5 ______
3. 38 + 5 ______
4. 48 + 5 ______
5. $\begin{array}{r} 53 \\ -\ 15 \\ \hline \end{array}$
6. 16 + ______ = 20
7. 6 + ______ = 20
8. 4 + ______ = 20
9. 14 + ______ = 20
10. $\begin{array}{r} 72 \\ -\ 24 \\ \hline \end{array}$

11.    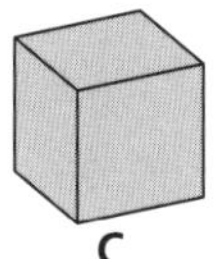 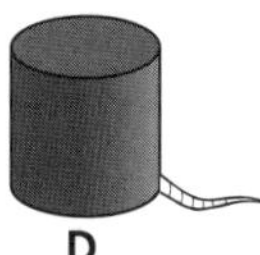

A B C D

a Which is a cylinder? ______

b Which is a cone? ______

12. Colour the change I would get from 50 cents when I spend 20 cents.

13. I get $3 pocket money each week.

How much will I get in 5 weeks? ______

14. What date is the:

a third Friday? ______

b second Sunday? ______

DECEMBER						
Sun	Mon	Tue	Wed	Thu	Fri	Sat
		1	2	3	4	5
6	7	8	9	10	11	12
13	14	15	16	17	18	19
20	21	22	23	24	25	26
27	28	29	30	31		

37:2 out of 14

1. 16 + 9 ______
2. 26 + 9 ______
3. 36 + 9 ______
4. 46 + 9 ______
5. $\begin{array}{r} 72 \\ -\ 39 \\ \hline \end{array}$
6. 44 + ______ = 60
7. 47 + ______ = 60
8. 64 + ______ = 80
9. 35 + ______ = 70
10. $\begin{array}{r} 54 \\ -\ 18 \\ \hline \end{array}$

11. I have $1.20. I need $2.40 to buy an ice-cream.

How much more do I need? ______

12. Colour the change I would get from $2 when I spend $1.45. ______

13. 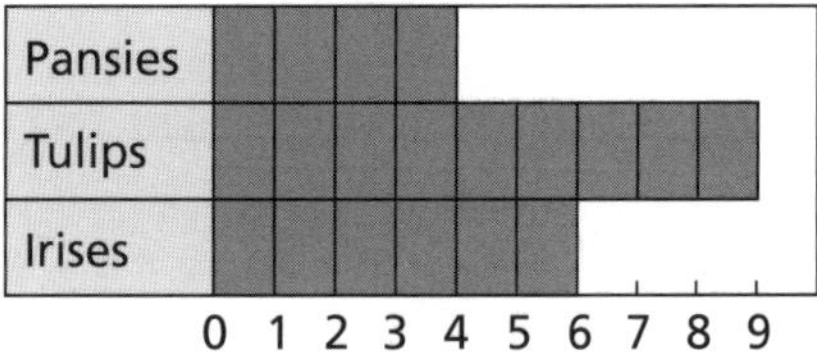

a Which was the smallest group? ______

b How many irises and tulips? ______

c How many more tulips than pansies? ______

14. a 600 000 + 3000 + 70 + 9 = ______

b 900 000 + 50 + 6000 + 1 = ______

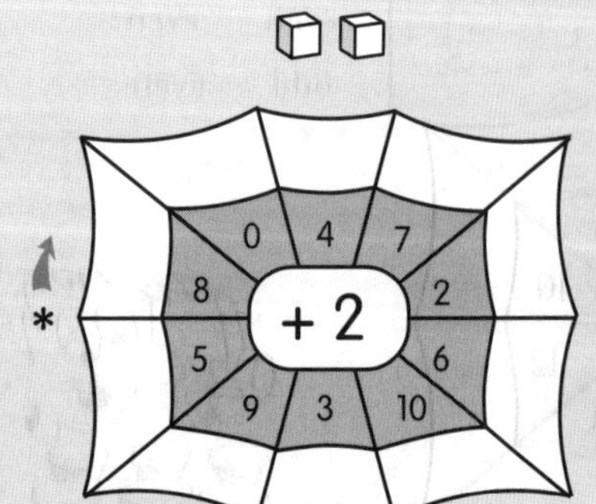

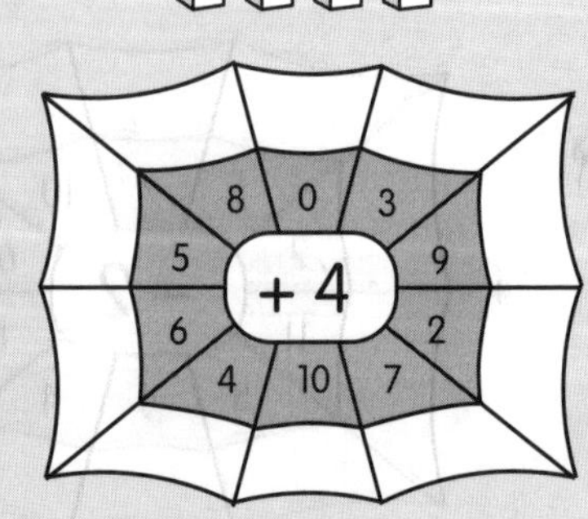

37:3 ☐ out of 6

1.

tens	ones
4	5
+3	2

2.

tens	ones
6	2
−3	1

3. Match each fraction to its decimal. ______

$\frac{2}{10}$	0·1
$\frac{7}{10}$	0·7
$\frac{1}{10}$	0·2

$1\frac{8}{10}$	1·6
$1\frac{6}{10}$	1·3
$1\frac{3}{10}$	1·8

4. Colour the change I would get from $2 when I spend 65 cents.

5.

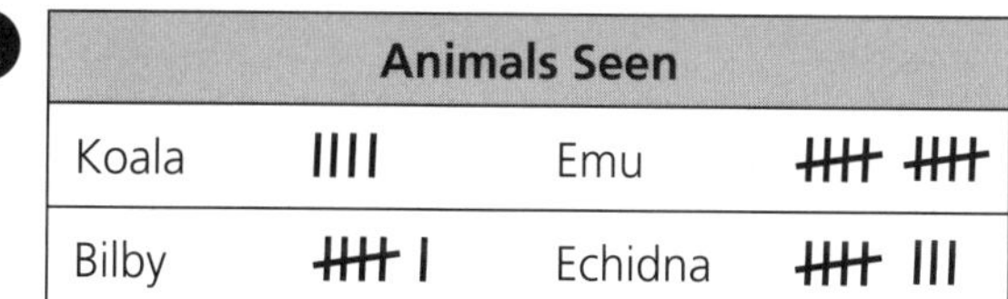

Animals Seen			
Koala	IIII	Emu	~~IIII~~ ~~IIII~~
Bilby	~~IIII~~ I	Echidna	~~IIII~~ III

a How many bilbies were seen? ______

b Which animal was seen most often? ______

c How many more echidnas than koalas were seen? ______

6. Jack caught 17 fish and Scott caught 6 more than Jack. How many did they catch altogether? ______

37:4 Extension

☐ out of 8

1. How many 50c coins make $5? ______
2. How many 20c coins make $5? ______
3. How many 10c coins make $5? ______
4. How many 5c coins make $5? ______
5. We bought 67 apples and ate 34, then I was given 12 more. How many do we have now? ______
6. Circle three quarters of the boats.

7. Add 15 each time.

30, ______, ______, ______, ______, ______

8. Ten toy soldiers were standing in a row. Between each pair there are two toy cars. How many toys are there altogether? ______

Challenge

Write down each set of 3 counting numbers that have a total of 7.
(Numbers can be used more than once.)

______

______

______

______

Fill out this table for the person measured in unit 1.

Name: ______ **Date:** ______

Age: ______	Mass: ______ kg	Shoe size: ______
Height: ______ cm	Waist: ______ cm	Neck size: ______ cm

How have these measurements changed since unit 1 was done?

______

How would you name the lizards?

Find two lizards hidden on each page of this book.

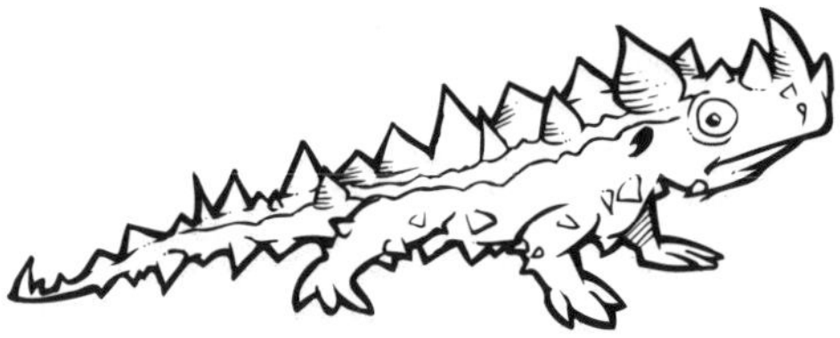

1 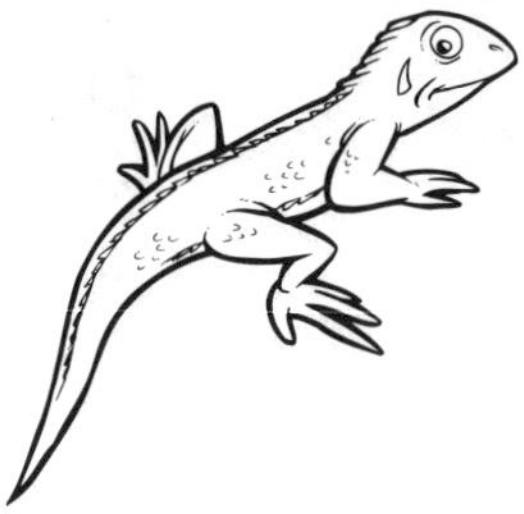 ____ 2 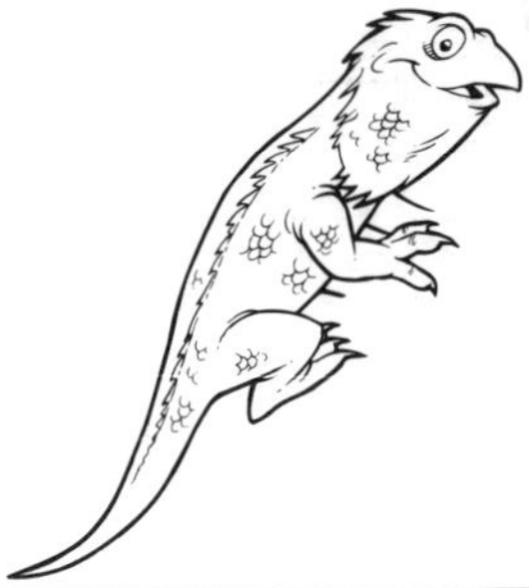 ____ 3 ____

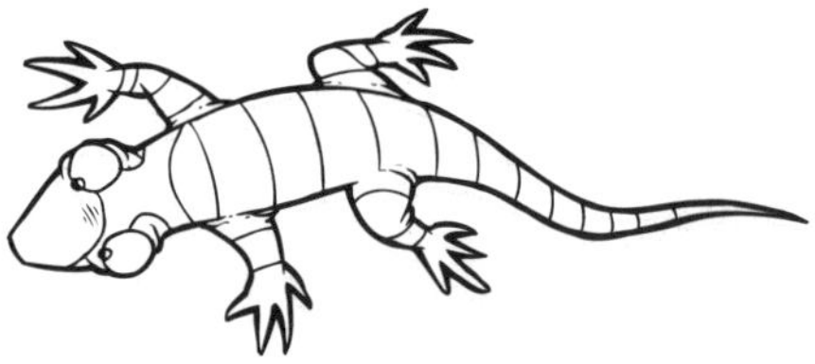

4 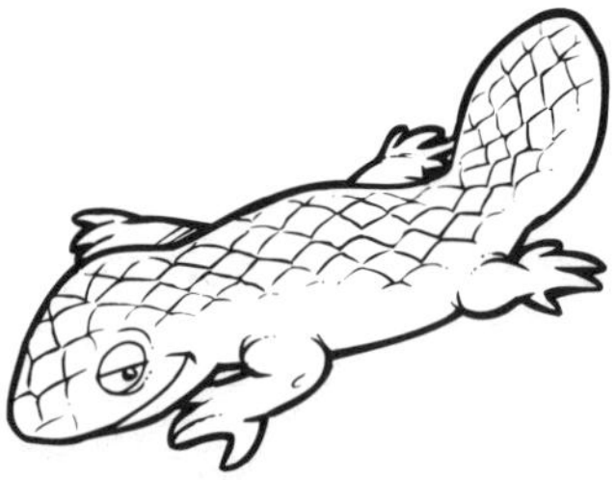 ____ 5 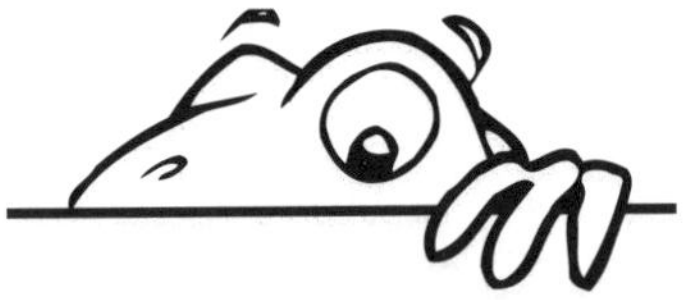 ____ 6 ____

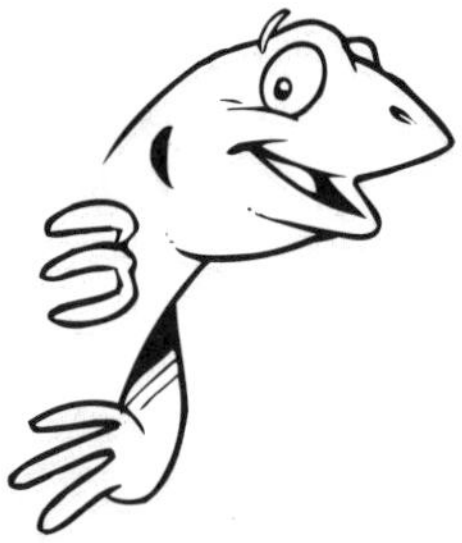

7 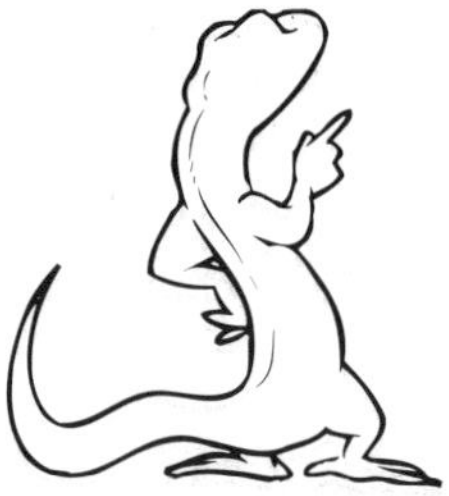 ____ 8 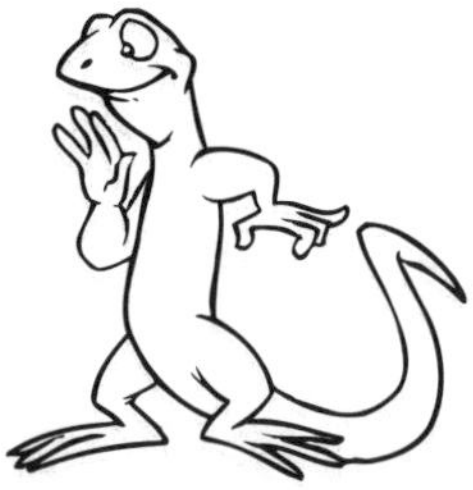 ____ 9 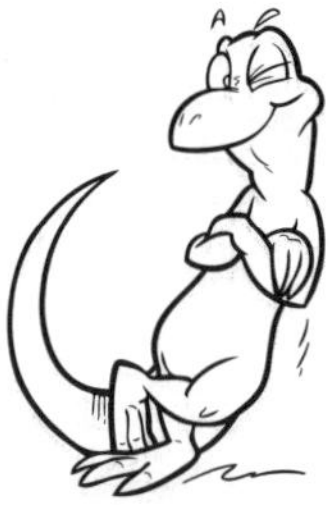 ____ 10 ____

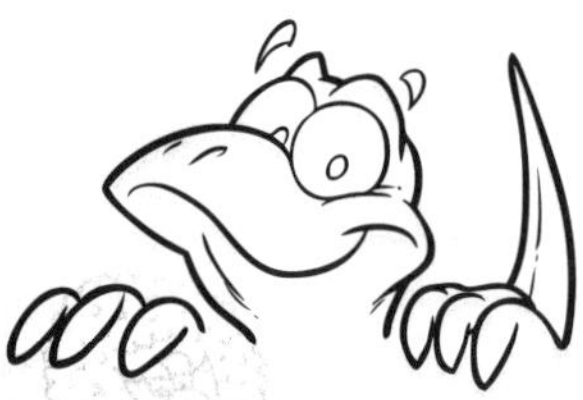

11 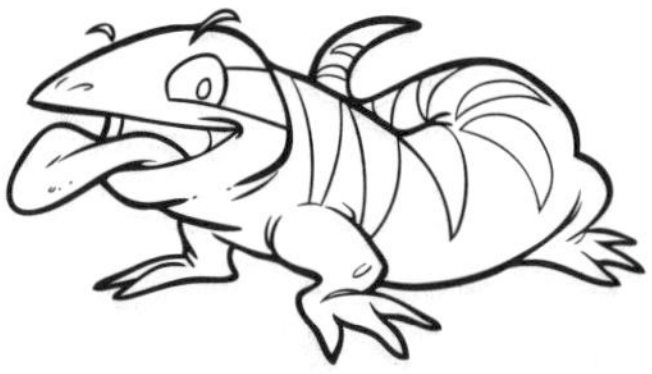 ____ 12 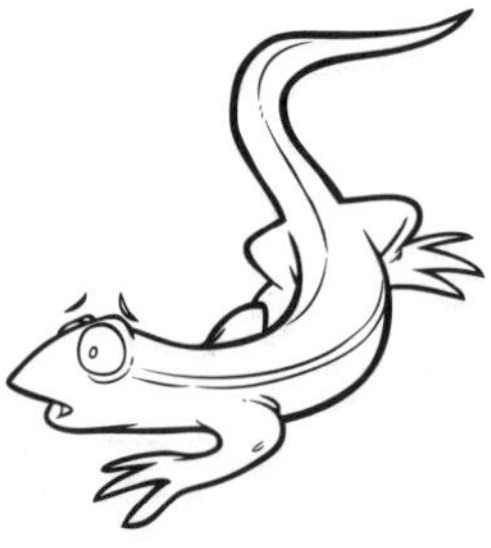 ____ 13 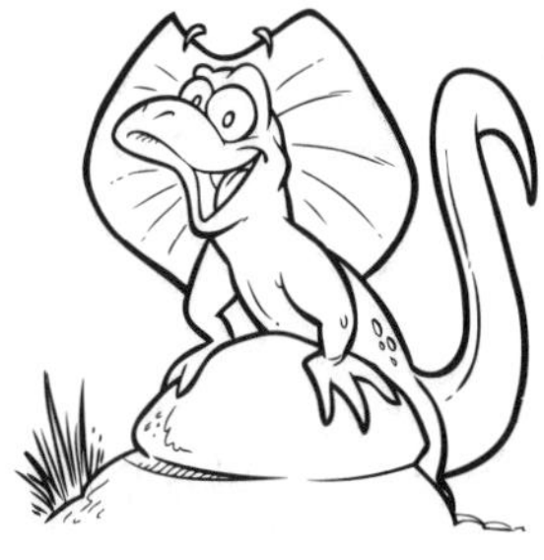 ____ 14 ____

Real names:

1 Horned lizard **2–3** Bearded dragon **4** Gecko **5** Shingle-back lizard

6–11 Common garden lizard **12** Blue-tongue lizard **13** Skink **14** Frilled-neck lizard